SpringerBriefs in Computer Science

Series Editors
Stan Zdonik
Shashi Shekhar
Jonathan Katz
Xindong Wu
Lakhmi C. Jain
David Padua
Xuemin (Sherman) Shen
Borko Furht
V.S. Subrahmanian
Martial Hebert
Katsushi Ikeuchi
Bruno Siciliano
Sushil Jajodia
Newton Lee

More information about this series at http://www.springer.com/series/10028

Eric Hardin • Helena Mitasova • Laura Tateosian
Margery Overton

GIS-based Analysis of Coastal Lidar Time-Series

Springer

Eric Hardin
Department of Physics
North Carolina State University
Raleigh, NC, USA

Helena Mitasova
Department of Marine, Earth
  and Atmospheric Sciences
North Carolina State University
Raleigh, NC, USA

Laura Tateosian
Center for Geospatial Analytics
North Carolina State University
Raleigh, NC, USA

Margery Overton
Department of Civil, Construction
  and Environmental Engineering
North Carolina State University
Raleigh, NC, USA

ISSN 2191-5768   ISSN 2191-5776 (electronic)
ISBN 978-1-4939-1834-8   ISBN 978-1-4939-1835-5 (eBook)
DOI 10.1007/978-1-4939-1835-5
Springer New York Heidelberg Dordrecht London

Library of Congress Control Number: 2014947349

Printed on acid-free paper

Springer is part of Springer Science+Business Media (www.springer.com)

Contents

Chapter 1
Introduction

Management of highly dynamic coastal landscapes requires repeated mapping and analysis of observed changes. Modern mapping techniques such as lidar increased the frequency and level of detail in coastal surveys and new methods were developed to extract valuable information from these data using Geographic Information Systems. In this chapter we discuss mapping of coastal change, on-line data resources, and the basics of installation and working with open source GRASS (Geographical Resources Analysis Support System) GIS used in this book.

1.1 Mapping Coastal Terrain Change

The present day coastal landscape is the result of complex interactions between natural processes and anthropogenic activities. Rapid urban development combined with increased shore erosion and severe storm impacts create new challenges for coastal management (Fig. 1.1). Quantification, modeling, and visualization of short term evolution of coastal systems is needed to better understand the impacts of natural processes and anthropogenic interventions. Identification of areas susceptible to high rates of erosion, accurate mapping of elevation and sand volume change and assessment of coastal vulnerability due to storm surge is critical for responsible coastal planning and management (Stockdon et al. 2007).

Numerous studies have demonstrated advantages of lidar surveys for assessment of shoreline and dune erosion (Burroughs and Tebbens 2008; Overton et al. 2006; Sallenger Jr et al. 2003; Stockdon et al. 2002). Lidar-based, bare earth Digital Elevation Models (DEMs) have been widely used for quantification of beach and dune volume change (Mitasova et al. 2004; Overton et al. 2006; White and Wang 2003), including assessment of major storm and hurricane impacts (Sallenger et al. 2006). The high density of lidar data points and near-annual frequency of coastal mapping in some regions provide time series of elevation data that can be used

© The Author(s) 2014   1
E. Hardin et al., *GIS-based Analysis of Coastal Lidar Time-Series*, SpringerBriefs
in Computer Science, DOI 10.1007/978-1-4939-1835-5_1

Fig. 1.1 Coastal management challenges on North Carolina Outer Banks: (**a**) storm impacts in Rodanthe (Hurricane Sandy, NCDOT 2012); (**b**) beach erosion in Nags Head (Hurricane Isabel, USGS 2003); (**c**) sand transport threatens homes and infrastructure (Nor'easter Athena, NCDOT 2012)

to extract new information about spatial patterns of coastal dynamics using raster and feature-based techniques. The changes in lidar technology over the past decade produced data sets with different accuracies, scanning patterns, and point densities. For this reason, geospatial analysis, when applied to multi-year lidar time series, also needs to address the issues of accurate data integration and computation of a consistent set of elevation models. Advanced three-dimensional Geographic Information Systems (GIS) provide a means for efficient integration of these new types of measurements. Once this integration is complete, GIS can be used to perform a wide range of sophisticated analyses and visualizations (Mitasova et al. 2011). This book explains both the necessary preprocessing and the subsequent analysis accompanied by step by step instructions and scripts applied to data sets from the North Carolina coast.

Lidar data and imagery for the coastal United States can be downloaded from the "Digital Coast", a National Oceanic and Atmospheric Administration operated website (National Oceanic and Atmospheric Administration Coastal Services Center 2010). The website provides tools for searching and pre-processing of data, such as coordinate transformation and gridding. It also allows users to select a wide range of data types and formats, such as all return, first return or bare ground points in the `las/` format or an `ascii/` text file. In this book, we use data for the coast of North

Carolina (NC) downloaded from the "Digital Coast". Additional data, including extensive collections of aerial imagery for NC are available from the NC Department of Transportation (NC DOT). High accuracy NC DOT benchmarks measured along the centerline of the highway NC-12 can be downloaded at http://www.obtf.org/NC12Alignment/NC12.htm. These benchmarks can be used to identify and reduce systematic error in the lidar.

1.2 GRASS GIS and Sample Data Set

The examples in this book process and analyze coastal lidar time series using the Geographic Resources Analysis and Support System (GRASS)—the free and open source GIS, specifically the GRASS7.0 release. The software is available to download for free from http://grass.osgeo.org/. The easiest to start with are the pre-compiled binary packages with installers available for Linux, MS-Windows, and Mac OS X. The basic terminology and data organization in GRASS7 is described in the GRASS GIS Quickstart document (http://grass.osgeo.org/grass71/manuals/helptext.html).

After installing the GRASS software create a directory where you will store all GRASS data. Name this directory grassdata/. This directory is often referred to as GIS data directory or GISDBASE. Within GISDBASE, GRASS data are organized into projects called LOCATIONS, which are defined by their coordinate system and spatial extent. When GRASS is started for the first time, you will be provided an option to navigate to and choose to work within an existing LOCATION, or define a new LOCATION using the Location wizard. LOCATIONS are subdivided into MAPSETS, which are used to organize data for sub-projects or for different users. Each LOCATION has a MAPSET called PERMANENT which is used for storing the coordinate system information and baseline geospatial data for the given project.

You can find all data sets used in this book at http://geospatial.ncsu.edu/osgeorel/data.html. Before starting GRASS, download the data set northcarolina_coast_spm.zip and unpack it in your grassdata/ directory. The data set is provided as a LOCATION which includes North Carolina boundaries in its PERMANENT MAPSET and a NagsHead_series/ MAPSET with time series of lidar-derived DEMs. The DEMs represent coastal topography along 1 km of shoreline at 1 m resolution in the town of Nags Head, NC, next to Jockeys Ridge State Park (Fig. 1.2). The time series contains series of time snapshots starting in 1996 (Mitasova et al. 2010). Additional data used in this book can also be downloaded from this website. These include the point cloud series for Jockey's ridge and Rodanthe (JR_*_lidar.txt and R_*_lidar.txt respectively), the road centerline (road_centerline.txt), and the road surface point cloud (DARE_BE*.txt).

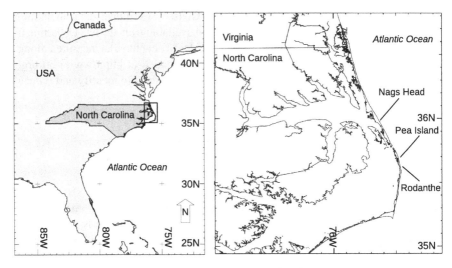

Fig. 1.2 Location of the study sites on Outer Banks, North Carolina

Once GRASS is installed and the data set downloaded and unpacked in the grassdata/ directory, you can start GRASS by clicking on its icon or from the terminal by typing

```
grass70
```

A start-up menu will open (Fig. 1.3) and you will be asked to select the GIS Data Directory (grassdata/), LOCATION (northCarolina_coast_spm/), and MAPSET (NagsHead_series/). After starting GRASS, a welcome message appears in the terminal and the Graphical User Interface (GUI) with the Map Display window and GIS Layers Manager will open.

If you are not familiar with GRASS you can get started by following a video tutorial at http://courses.ncsu.edu/mea582/common/media/02/getting_started_GUI_1. mov or check out the latest GRASS videos on YouTube https://www.youtube.com/ results?search_query=grass+gis. The GRASS Reference manual gives a description of all GRASS commands and is available online at http://grass.osgeo.org/grass70/ manuals/index.html.

You can perform GRASS GIS analysis using the graphical user interface (GUI), or using commands that correspond to the GUI tools and can be run in the GRASS shell or in the 'Command console'. GRASS workflows can also be executed via Python code. A Python interpreter is embedded within the GRASS GIS software. The easiest approach for running GRASS commands with Python is to run these scripts inside of GRASS.

Examples in this book are written as command line workflows (equivalent to shell code), which are executed in the GRASS shell or the command console. Examples using Python scripting are provided when more complicated control structures or string parsing are needed. The Python code should be executed using the Python

Fig. 1.3 GRASS startup screen

interpreter inside the GRASS environment. Note, that with the default settings, the Python scripts can not be executed outside of the GRASS environment.

1.3 Organization of This Book

This book focuses on GIS-based processing, analysis, and visualization of coastal lidar time-series. The descriptions of the approaches outlined here are accompanied by examples, which are implemented using GRASS GIS, and freely available sample data.

The next chapter describes the initial data processing that is necessary to integrate the lidar data into a consistent raster time series. The following two chapters explain per-cell statistical analysis and techniques for extracting coastal features (including shorelines, dunes, and structures). Chapter 5 covers volumetric analysis (i.e., volume estimation and change-based metrics). Chapter 6 provides an introduction to techniques for coastal data visualization and explains visualization in space-time cube. The appendix includes a summary of data and color tables for raster maps used throughout the book.

References

Burroughs, S. and Tebbens, S. (2008). Dune retreat and shoreline change on the Outer Banks of North Carolina. *Journal of Coastal Research*, 24:104–112. DOI: 10.2112/05-0583.1.

Mitasova, H., Drake, T., Bernstein, D., and Harmon, R. (2004). Quantifying rapid changes in coastal topography using modern mapping techniques and geographic information system. *Environmental and Engineering Geoscience*, 10:1–11. DOI: 10.2113/10.1.1.

Mitasova, H., Hardin, E., Overton, M., and Kurum, M. (2010). Geospatial analysis of vulnerable beach-foredune systems from decadal time series of lidar data. *Journal of Coastal Conservation*, 14:161–172. DOI: 10.1007/s11852-010-0088-1.

Mitasova, H., Hardin, E., Starek, M., Harmon, R., and Overton, M. (2011). Landscape dynamics from LiDAR data time series. *Geomorphometry 2011, Redlands, CA*, pages 3–6.

National Oceanic and Atmospheric Administration Coastal Services Center (2010). NOAA Coastal Services Center Coastal Lidar. http://csc.noaa.gov/digitalcoast/dataregistry/#/ Accessed 16 Jun. 2014.

Overton, M., Mitasova, H., Recalde, J., and Vanderbeke, N. (2006). Morphological evolution of a shoreline on a decadal time scale. *Proceedings of the 30th International Conference on Coastal Engineering, San Diego, California*, page 3851.

Sallenger, A., Stockdon, H., Fauver, L., Hansen, M., Thompson, D., Wright, C., and Lillycrop, J. (2006). Hurricanes 2004: An overview of their characteristics and coastal change. *Estuaries and Coasts*, 29:880–888. DOI: 10.1007/BF02798647.

Sallenger Jr, A., Krabill, W., Swift, R., Brock, J., List, J., Hansen, M., Holman, R., Manizade, S., Sontag, J., Meredith, A., et al. (2003). Evaluation of airborne topographic lidar for quantifying beach changes. *Journal of Coastal Research*, 19(1):125–133. ISSN: 0749-0208.

Stockdon, H., Sallenger, A., and Holman, R. (2007). A simple model for the spatially-variable coastal response to hurricanes. *Marine Geology*, 238:1–20. DOI: 10.1016/j.margeo.2006.11.004.

Stockdon, H., Sallenger, A., List, J., and Holman, R. (2002). Estimation of shoreline position and change from airborne topographic lidar data. *Journal of Coastal Research*, 18:502–513.

White, S. and Wang, Y. (2003). Utilizing DEMs derived from LIDAR data to analyze morphologic change in the North Carolina coastline. *Remote Sensing of Environment*, 85(1):39–47. DOI: 10.1016/S0034-4257(02)00185-2.

Chapter 2
Processing Coastal Lidar Time Series

In this chapter, we analyze time series of lidar data point clouds to assess the point density, gaps in coverage, spatial extent and accuracy. Based on this analysis and a given application, we select an appropriate resolution and interpolation method for computating raster-based digital elevation models (DEM). We explain a per raster-cell average approach and two splines-based approaches for computating DEMs. Finally, we discuss how to assess systematic error using geodetic benchmarks or other ground truth point data and how to correct any shifted DEMs to create a consistent DEM time series.

2.1 General Workflow

Time series of lidar point clouds include data from multiple surveys often acquired for different purpose by various types of lidar technology. To understand the properties of point clouds in the time series, we first analyze the data at a sequence of resolutions and then apply interpolation to compute a DEM at the selected resolution. The methodology, which can be applied to both first return or bare ground data can be summarized as follows:

- Integrate the point-cloud data acquired from various sources within a single coordinate system.
- Perform per-cell statistical analysis of point data at a hierarchical set of resolutions, and use the results to select a common DEM resolution.
- Derive the spatial extent of each survey and a mask for the study area from preliminary low-resolution DEMs computed using the mean elevation value for each cell.

© The Author(s) 2014
E. Hardin et al., *GIS-based Analysis of Coastal Lidar Time-Series*, SpringerBriefs
in Computer Science, DOI 10.1007/978-1-4939-1835-5_2

Table 2.1 Characteristics of the lidar surveys based on the available metadata

Agency,* Dates	Lidar equipment	Published point density	Published Accuracy vertical/ horizontal (m))
NOAA/NASA/USGS October 19, 1996 September 1 and 26, 1997 September 7, 1998; post-Bonnie† September 9, 1999; post-Dennis† September 18, 1999; post-Floyd† October 6, 1999	Airborne topographic Mapper II	1pt/3m	0.15/2.00
NCDENR/FEMA/NCFMP February 2001	Leica Geosystems aeroscan	1pt/3m	0.20/2.00
NASA/USGS September 18, 2003 pre-Isabel† September 21, 2003 post-Isabel†	EAARL	1pt/3m	0.15/2.00
JALBTCX August 28, 2004 September 28, 2005, post-Ophelia†	Compact hydrographic Airborne rapid total Survey (charts)	1pt/1m	0.3/1.4
NOAA March 27, 2008	IOCM	1pt/1m	0.3/1.4
NASA, USGS December 1, 2009, post-Nor'Ida†	EAARL	1pt/1m	0.2/0.75
NOAA August 8, 2011, post-Irene†		1pt/1m	0.3/1.4

*NASA = National Aeronautics and Space Administration, NOAA = National Oceanic and Atmospheric Administration, USGS = U.S. Geological Survey, NCDENR = North Carolina Department of Environment and Natural Resources, FEMA = Federal Emergency Management Agency, NCFMP = North Carolina Floodplain Mapping Program, JALBTCX = Joint Airborne Lidar Bathymetry Center of Expertise, EAARL = Experimental Advanced Airborne Research Lidar, IOCM = Integrated Ocean and Coastal Mapping
† Hurricane names

- Compute more detailed, smoothed DEMs for the masked study area using spatial interpolation.
- Compare the DEMs with high accuracy ground-based data to remove potential systematic errors and verify the accuracy of each DEM.

The result of this procedure is a consistent series of DEMs which have a common resolution and are clipped to a common spatial extent. To illustrate the workflow we use the provided series of lidar point clouds acquired along the coast of NC since 1996 (Table 2.1). The published horizontal accuracy of this data is 2 m, while the vertical accuracy is 0.12–0.20 m.

2.2 Analysis of Lidar Point Clouds

Data from lidar surveys acquired over the span of several years and for a wide range of applications have varied point densities, spatial extents, and accuracy. We use point per-cell statistics to map the distribution of point densities (as the number of points found in each raster cell) and the range of values in a raster cell. We use this to create a low resolution DEM by computing mean point elevation per cell. This information is helpful when selecting a common resolution for the entire series of DEMs.

To compute the per cell statistics we first set the resolution using g.region and then import the lidar points from a lidar text files. In the code below, we use the r.in.xyz GRASS command to compute the point count for each raster cell at a resolution 5 m. Then we use the r.univar module to calculate mean point count per raster cell.

```
# Purpose: Get mean cell statistics.
# Refer to Ch. 1 Sec. 1.2 for information on where to
# download the data.
# Start grass with location northcarolina_coast_spm
# and mapset NagsHead_series.
grass70

# Set region to Jockey's Ridge area and display
# provided DEM in 2D and 3D
g.region rast=NagsHead_series_1m -p
d.rast NH_2008_1m

# Compute number of lidar points per grid cell at 5m
# resolution for the 1999 survey
g.region res=5
r.in.xyz input=JR_19990909.txt output=JR_stats_n \
    method=n fs=','
r.null map=JR_stats_n setnull=0
# Get the mean per cell count
r.univar -ge map=JR_stats_n
```

To perform this kind of analysis for a set of resolutions and series of lidar point clouds, we use Python code. In the code below, we compute the point count, elevation range, and mean elevation for each raster cell at the resolutions of 0.5, 2, 5, and 10 m. Then we calculate global statistics on the range raster maps for each resolution.

```
# Purpose: Get mean cell statistics.
import grass.script as grass
files = ['JR_19971002.txt', 'JR_19990909.txt',
    'JR_2001.txt', 'JR_20051126.txt', 'JR_20080327.txt'
    ]
```

```
resolutions=[ 0.5, 2, 5, 10 ]
grass.run_command('g.region',
    region='NagsHead_series_1m')
report = 'date\tres\tn\trange\n'
for f in files:
    report += f + '\n'
    for res in resolutions:
        report += '\t' + str(res) + '\t'
        # Set the resolution.
        grass.run_command('g.region', res=res)
        # Get the per cell count.
        grass.run_command('r.in.xyz', input=f,
            output='JR_stats_n', method='n', fs=',',
            overwrite=True)
        grass.run_command('r.null', map='JR_stats_n',
            setnull=0)
        # Get the mean per cell count.
        stats = grass.parse_command('r.univar',
            flags='ge', map='JR_stats_n')
        report += str(stats['mean']) + '\t'
        # Get the per cell range.
        grass.run_command('r.in.xyz', input=f,
            output='JR_stats_range', method='range',
            fs=',', overwrite=True)
        grass.run_command('r.mapcalc',
            expression='JR_stats_range_c=if(isnull(JR_
                stats_n),
        null(), JR_stats_range)', overwrite=True)
        # Get the mean per cell range.
        stats = grass.parse_command('r.univar',
            flags='ge', map='JR_stats_range_c')
        report += str(stats['mean']) + '\n'
grass.run_command('g.region',
    region='NagsHead_series_1m')
print( report )
```

The results of this lidar point cloud analysis at a hierarchy of resolutions for selected surveys are summarized in Table 2.2 and illustrated by Fig. 2.1. At a resolution of 10 m, the mean range of elevations within the raster cells exceeds 1 m for all surveys and 2 m for the last three surveys, indicating that important features may be lost at this resolution. At a resolution of 2 m, the mean range is between 0.08 and 0.65 m and the number of points per raster cell is less than one for older surveys, indicating the need for interpolation. At 0.5 m resolution, the within-cell mean range

Table 2.2 Mean per cell
point count and elevation
range at 0.5, 2 and 10 m
resolution for selected lidar
surveys of Nags Head

	Grid size (m)	Points per cell	Range (m)
1997	0.5	1.102	0.020
	2	2.559	0.249
	10	45.522	1.753
1999	0.5	1.113	0.023
	2	3.295	0.315
	10	60.012	1.822
2001	0.5	1.000	0.000
	2	1.006	0.011
	10	7.394	1.358
2005	0.5	1.267	0.034
	2	6.025	0.560
	10	145.361	2.669
2008	0.5	1.030	0.012
	2	3.589	0.444
	10	85.303	2.143

was less than the published data accuracy and interpolation is necessary for all
surveys. To preserve the shape of the buildings, we select 0.5 m resolution and the
time series of DEMs will be created by interpolation.

2.3 Computing DEMs

To compute a consistent series of DEMs we first derive masks of mapped areas for
each survey, then we apply interpolation using the method most appropriate for our
application.

2.3.1 Masking Surveyed Areas

Interpolating lidar point data to high resolution DEMs is only meaningful in regions
with adequate point coverage, We can mask out low density point regions so that
only high density regions are interpolated. Masking is also important because it can
substantially reduce the processing time during data analysis. We identify regions
to mask by first importing the lidar points at a resolution much greater than the lidar
point space (The high resolution value is selected based on the point density analysis
in Sec. 2.2). Then we set each cell in the resultant raster to 1 if the cell contains any
lidar data points or a 'no-data' value if it does not (Fig. 2.2). The following GRASS
code sets the resolution to 5 m and uses the `r.in.xyz` and `r.mapcalc` functions
to perform these two steps and create a mask based on point density:

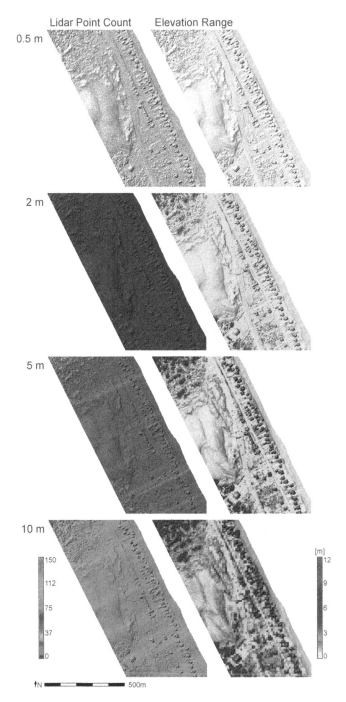

Fig. 2.1 Point density (lidar point count) and elevation range at different resolutions

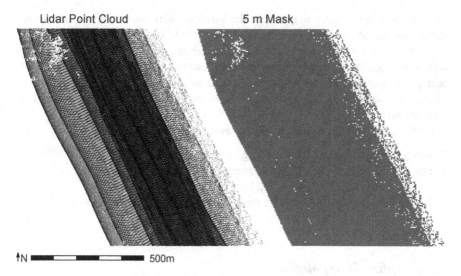

Fig. 2.2 Point cloud and a derived mask based on 1999 lidar

```
# Purpose: Create a masked survey area.
g.region NagsHead_series_1m res=5
r.in.xyz input=JR_1999.txt output=JR_1999_n_5m \
    method=n fs=',' --o
r.mapcalc expression='JR_1999_mask=if( JR_1999_n_5m \
    == 0, null(), 1 )' --o
```

Raster operations (including interpolation) can then be limited to the mask by running the following commands:

```
# Purpose: Limit raster operations with a mask.
g.region res=0.5
r.mask input=JR_1999_mask
```

Raster operations will continue to be limited to the mask area until the mask is removed by running r.mask with the -r flag:

```
# Purpose: Remove raster mask.
r.mask -r input=JR_1999_mask
```

With the mask set up we can now interpolate the DEMs using a method suitable for the given application.

2.3.2 Binning

When a lidar point cloud is available in an ASCII text format (such as x,y,z tuples) and has at least one point in each raster cell at a fixed resolution, a DEM surface can be generated directly from the lidar points using the r.in.xyz module. The module computes a raster map where the value in each raster cell is a univariate statistic of the lidar data points contained in that cell. For this reason, the method is referred to as binning. The `method` parameter specifies the statistical measure, such as the maximum, minimum, or mean elevation value. We use the analysis of lidar point density outlined in the Sec. 2.2 to select the adequate resolution. For binning, a resolution of 2 m was chosen to ensure that most grid cells contained at least one lidar point.

DEMs are usually computed by setting `method` to `mean` (Fig. 2.3a).

```
# Purpose: Create DEM using raster statistics.
g.region region=NagsHead_series_1m res=2
r.in.xyz input=JR_20080327.txt \
    output=JR_20080327_binmean1m method=mean fs=','
```

2.3.3 Spline Interpolation

Continuous DEMs at resolutions higher than the average point spacing can be computed using spatial interpolation. GRASS7 provides two spline-based modules for bivariate interpolation: `v.surf.rst` and `v.surf.bspline`.

Detailed, smoothed sets of DEMs and topographic parameters (slope, aspect and curvatures) can be computed using the regularized spline with tension (RST) method (Mitasova et al. 2005). RST belongs to interpolation functions that minimize the deviations from the measured points and a smoothness seminorm (Mitas and Mitasova 1999). The RST smoothness seminorm includes derivatives of all orders with their weights decreasing with the increasing derivative order leading to the following function:

$$z(\mathbf{r}) = a_1 + \sum_{j=1}^{N} \lambda_j R(\varrho_j) \tag{2.1}$$

$$R(\varrho_j) = -[E_1(\varrho_j) + \ln(\varrho_j) + C_E] \tag{2.2}$$

where $z(\mathbf{r})$ is elevation at a point $\mathbf{r} = (x, y)$, a_1 is a trend, λ_j are coefficients, N is the number of given points, $R(\varrho_j)$ is a radial basis function, $\varrho_j = (\varphi r_j/2)^2$, φ is a generalized tension parameter, $r_j = |\mathbf{r} - \mathbf{r}_j|$ is a distance, $C_E = 0.577215$ is the Euler constant, and $E_1(\varrho_j)$ is the exponential integral function (Abramowitz and Stegun 1965; Mitášová and Mitáš 1993). The coefficients a_1 and $\{\lambda_j\}$ are obtained by solving the system of linear equations:

$$\sum_{j=1}^{N} \lambda_j = 0. \tag{2.3}$$

$$a_1 + \sum_{j=1}^{N} \lambda_j \left[R(\varrho_j) + \delta \frac{w_0}{w_j} \right] = z(\mathbf{r}_i), \qquad i = 1, \ldots, N \qquad (2.4)$$

where w_0/w_j are positive weighting factors representing a smoothing parameter at each given point $\mathbf{r_j} = (x_j, y_j)$.

The method has both geostatistical and physical interpretation (Mitas and Mitasova 1999). It is formally equivalent to universal kriging with the choice of the covariance function determined by the smoothness seminorm. The intuitive physical interpretation of this method is a thin surface that can be tuned from a rigid plate to a rubber sheet by changing its tension (Fig. 2.3). The tension parameter φ controls the distance over which the given points influence the resulting surface while smoothing controls the vertical deviation of the surface from data points. By using an appropriate combination of tension and smoothing, it is possible to apply the function to various types of surfaces from smoothly changing topography to rough terrain, and select a level of detail represented by a DEM without changing the resolution. The optimal values of parameters can often be found by minimizing the cross validation error (Hofierka et al. 2002; Mitas and Mitasova 1999). The tension and smoothing parameters for each DEM computation can be optimized to reduce the noise and ensure a comparable level of detail in each DEM (see Mitasova et al. (2005), or Neteler and Mitasova (2008) for more details on RST implementation and optimization of its parameters for lidar data).

The RST interpolation for the entire DEM series along with computation of topographic parameters (slope, aspect, profile and tangential curvatures) can be carried out in GRASS by importing the lidar data points using the v.in.ascii function and then interpolating the points using the v.surf.rst function, as in the following Python script:

```
# Purpose: Import point clouds and interpolate using
# the RST method.
# Refer to Ch. 1 Sec. 1.2 for information on where to
# download the data.
import grass.script as grass

# Find and set region from point cloud.
grass.run_command( 'v.in.ascii',
    input='R_19961016_lidar.txt',
  output='temp', format='point', separator=',',
  skip=0, x=1, y=2, z=3)
grass.run_command( 'g.region', flags='pa',
    vect='temp', res=0.5 )

# Import and interpolate point clouds.
dates = [19961016, 19971002, 19980907, 19990909,
    19990918, 19991104, 2001, 20030916, 20030921,
```

```
     20040925, 20051126, 20080327]
ten = [1200, 1200, 500, 1000, 1000, 1000, 1500, 1000,
    1000, 1500, 2000, 2000]
for i in range( len(dates) ):
    fin = 'R_'+str(dates[i])+'_lidar.txt'
    vect = 'R_'+str(dates[i])
    # Import lidar points that fall within the current
    # region.
    grass.run_command( 'v.in.ascii', flags='tbr',
        input=fin, output=vect, format='point',
        fs=',', skip=0, x=1, y=2, z=3)
    rast = 'R_'+str(dates[i])+'_05mrst'
    # Interpolate using RST with scale dependent
    # tension.
    grass.run_command( 'v.surf.rst', flags='tz',
        input=vect, elev=rast, slope=rast+'_slp',
        pcurv=rast+'_pcurv', tcurv=rast+'_tcurv',
        tension=ten[i], smooth=0.5, overwrite=True )
```

The value of the tension parameter is modified for each data set to account for the differences in point densities and level of detail. The script runs the RST interpolation with the -t flag, so that tension is not influenced by the data segmentation and normalization.

Another approach to generating smoothed sets of DEMs is bilinear or bicubic spline interpolation with Tykhonov regularization. In this approach each observation (or lidar data point) is interpreted as a linear combination of spline functions (Brovelli et al. 2004)

$$h_0(\underline{t}_m) = \sum_{lk} a_{lk} s_{\Delta^g}(\underline{t}_m - \tau_{lk}) + v_m \qquad (2.5)$$

where $h_0(\underline{t}_m)$ is the elevation of the mth lidar data point, $\underline{t}_m$ is the planimetric location of the lidar data point, a_{lk} is an unknown fitting parameter, s_{Δ^g} is an interpolation function with compact support (the range of which is described by Δ) and order g (e.g., $g = 1$ describes a bilinear function), τ_{lk} is the planimetric location of the spline interpolating function (which centers on raster cell lk), and v_m is an unobserved disturbance.

Equation (2.5) can be written in matrix form as

$$\underline{Y}_0 = A\underline{a} + \underline{v} \qquad (2.6)$$

where

$$\underline{Y}_0 = [\dots h_0(\underline{t}_m) \dots]^T \qquad (2.7)$$

$$\underline{a} = [\dots a_{lk} \dots]^T \qquad (2.8)$$

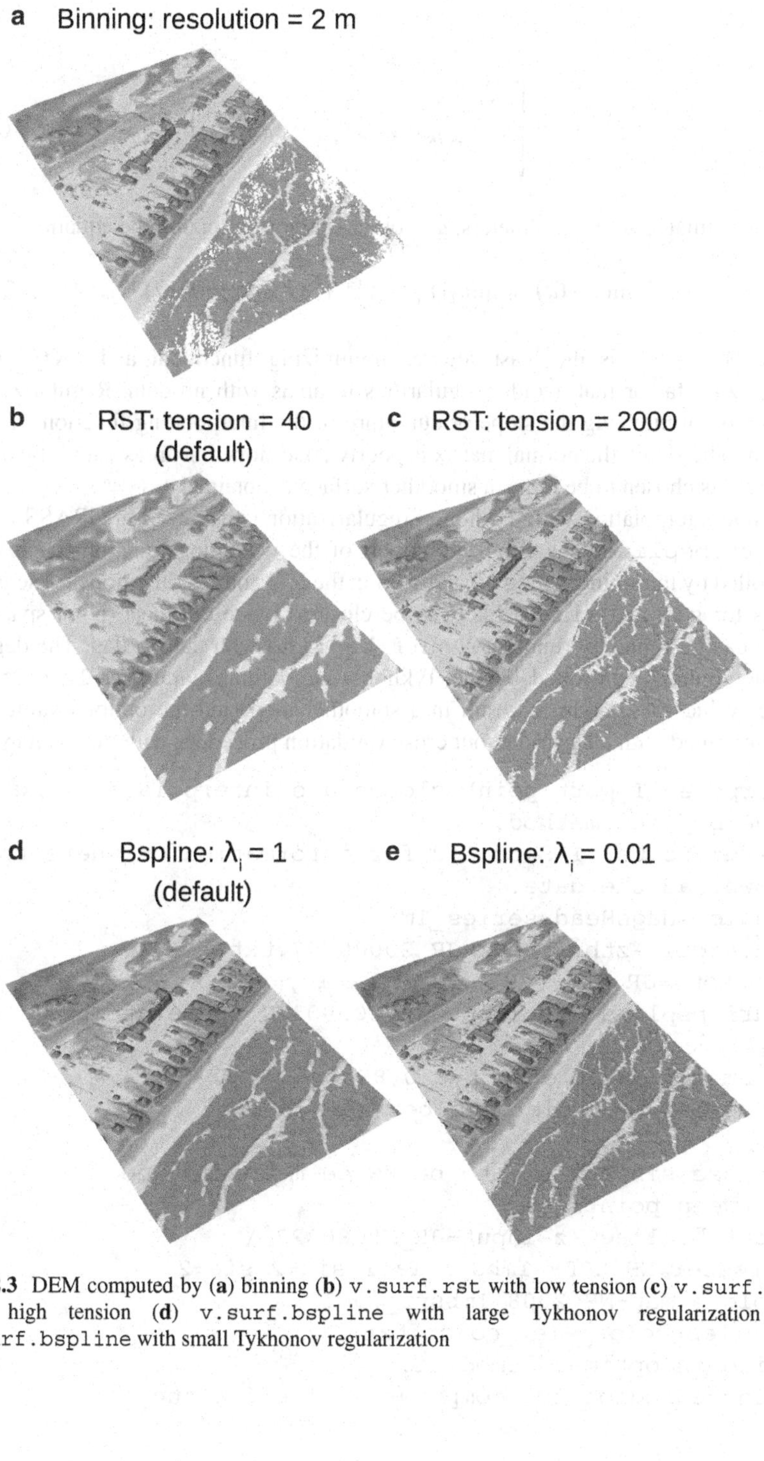

a Binning: resolution = 2 m

b RST: tension = 40 (default)   **c** RST: tension = 2000

d Bspline: $\lambda_i = 1$ (default)   **e** Bspline: $\lambda_i = 0.01$

Fig. 2.3 DEM computed by (**a**) binning (**b**) v.surf.rst with low tension (**c**) v.surf.rst with high tension (**d**) v.surf.bspline with large Tykhonov regularization (**e**) v.surf.bspline with small Tykhonov regularization

and

$$
A = \begin{bmatrix} \cdots & \cdots & \cdots \\ \cdots \sum_{lk} a_{lk} s_{\Delta^g} (\underline{t}_m - \tau_{lk}) \cdots \\ \cdots & \cdots & \cdots \end{bmatrix} \tag{2.9}
$$

The estimated set of parameters, $\hat{\underline{a}}$ is obtained by minimizing the equation

$$
\min \Psi(\underline{a}) = \min\{|\underline{Y}_0 - \underline{\hat{y}}|^2 + \lambda K(\underline{a})\} = \Psi(\underline{\hat{a}}) \tag{2.10}
$$

where $|\underline{Y}_0 - \underline{\hat{y}}|^2$ is the least squares minimizing functional and $\lambda K(\underline{a})$ is a regularizing factor that avoids singularities in areas with no data. Regularization is done by minimizing the slope or curvature of the interpolating function. If λ is chosen to be small, the normal matrix is poorly conditioned in areas with little or no data. If λ is chosen to be large, a smoother surface is obtained.

Spline interpolation with Tykhonov regularization is achieved in GRASS using v.surf.bspline. The compact support of the weighting function (i.e., Δ) is controlled by the spline step (sie and sin in the EW and NS directions). Adequate values for sie and sin are likely to be close to twice the mean point spacing, which can be found by running v.surf.bspline with the -e flag. The degree of smoothing is controlled by the Tykhonov smoothing parameter lambda_i. Larger values of lamda_i result in a smoother map, and the optimal value can be determined with a leave-one-out cross validation procedure with the -c flag.

```
# Purpose: Import point clouds and interpolate using
# the bspline method.
# Refer to Ch. 1 Sec. 1.2 for information on where to
# download the data.
g.region NagsHead_series_1m
v.in.ascii -ztbr input=JR_20080327.txt \
    output=JR_20080327 fs=',' x=1 y=2 z=3
v.surf.bspline -ze input=JR_20080327 raster=temp
[..]
 Estimated point density: 0.8537
 Estimated mean distance between points: 1.082
[..]
# Choose sin and sie to be twice mean distance
# between points.
v.surf.bspline -z input=JR_20080327 \
    raster=NH_2008_1mbspl_lam1 sin=2 sie=2
r.colors map=NH_2008_1mbspl_lam1 \
    rules=color_elev_coast.txt
# Find an optimal lambda_i.
# Reduce region for computational efficiency
```

```
g.region n=s+40 e=w+40
v.surf.bspline -c input=JR_20080327 \
    raster_output=temp sin=2 sie=2
g.region n=250670 s=249730 w=913366 e=914342 res=1
v.surf.bspline - input=JR_20080327 \
    raster_output=JR_2008_1mbspl_lam001 sin=2 sie=2 \
    lambda_i=0.01 --o
r.colors map=JR_2008_1mbspl_lam001 \
    rast=NH_2008_1mbspl_lam1
```

To interpolate the entire series, use the Python code above for the RST method but replace v.surf.rst with the command v.surf.bspline. DEMs resulting from large and small values of lamda_i are shown in Figs. 2.3 d and e.

2.4 Eliminating Water Surface Features

For many applications, such as volume calculations or shoreline extraction, elevation data representing water surface features should be set to 'no-data' values. After the DEMs are generated this can be achieved by setting elevations that are lower than the mean high water (MHW) elevation to 'no-data' values. Any remaining data regions that have a smaller area than the largest one are presumed to represent wave crests and other spurious data, so these are also set to the 'no-data' value (Fig. 2.4).

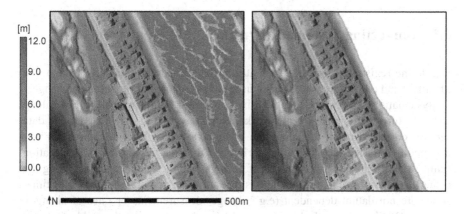

Fig. 2.4 Elimination of water surface features

```
# Purpose: Eliminate water surface features.
r.mapcalc \
expression='JR_20080327_05mbspl_ext_gt036=if
  (JR_20080327_05mbspl_ext>0.36,\
```

```
 1, null())' --o
r.to.vect input=JR_20080327_05mbspl_ext_gt036 \
   output=JR_20080327_05mbspl_ext_gt036 type=area --o

# Find the unique, database-generated category of the
# largest area.
# In this case, the category is 1.
v.report -s map=JR_20080327_05mbspl_ext_gt036 \
   option=area

v.extract input=JR_20080327_05mbspl_ext_gt036 \
   output=JR_20080327_05mbspl_ext_mask cats=1 --o
v.to.rast input=JR_20080327_05mbspl_ext_mask \
   output=JR_20080327_05mbspl_ext_mask use=val \
   value=1 --o
r.mapcalc\
  expression='JR_20080327_05mbspl_ext_masked=
  JR_20080327_05mbspl_ext\
  * float(JR_20080327_05mbspl_ext_mask)' --o

g.remove rast="JR_20080327_05mbspl_ext_gt036,
  JR_20080327_05mbspl_ext_mask"
g.remove vect="JR_20080327_05mbspl_ext_gt036,
  JR_20080327_05mbspl_ext_mask"
```

2.5   Correcting Systematic Errors

Due to the registration errors, lidar data can be shifted and this shift needs to be identified and corrected if the data are used for assessment of topographic change.

Systematic errors can be identified by comparing the interpolated DEMs along stable features and geodetic benchmarks in open areas (Fig 2.5). Our sample data set was corrected using the centerline of highway NC-12 because this road was not modified during the study time period and thus had a time-invariant elevation (unlike the erodible terrain surface). If no high-accuracy altimetric data along the centerline is available and if the metrics that are to be derived from the DEM time-series are not datum dependent (e.g., change measurements or rates of change), then the DEMs can simply be referenced to each other using the stable features, such as roads. Alternatively, if high-accuracy altimetric data is available, then for each lidar dataset, elevation differences between the high-accuracy data and lidar can be computed. The median difference quantifies the systematic error. Although mean and median errors are often comparable, the median is chosen for its lower sensitivity to outliers. In the relatively flat coastal terrain, systematic error can be assumed to be spatially constant and can be corrected by shifting the lidar-based DEMs so that the median difference becomes zero. Although the median error is

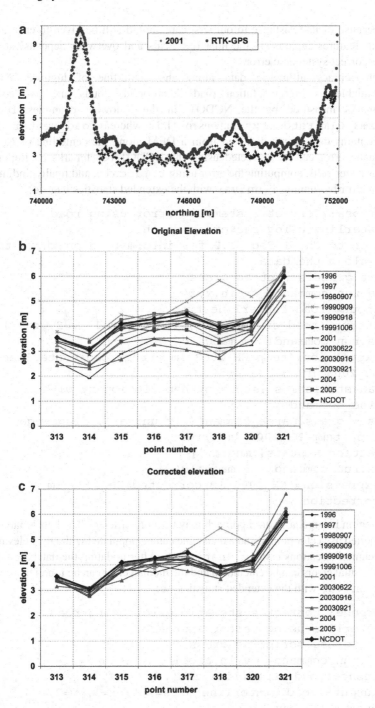

Fig. 2.5 (**a**) RTKGPS versus lidar profile along the road centerline. Elevation along the centerline of highway NC 12 from (**b**) uncorrected lidar and (**c**) lidar with corrected systematic error (Mitasova et al. 2009)

used because of its resistance to outliers, care should still be given to ensure that spurious features captured in the lidar (e.g., cars and overwash deposits) are not used to correct systematic error.

High-accuracy altimetric data along the centerline of highway NC-12 are available as high-resolution road lidar point clouds and as geodetic benchmarks measured by the NCDOT. In the following examples, DARE_ BE94zm3_01m_rstdm.txt contains road lidar, whereas road_centerline. txt contains data points digitized from a DEM along the centerline of NC-12. Systematic error can be corrected using the NCDOT benchmarks by importing them as raster cells, computing the error using r.mapcalc, and finally finding the median error by running r.univar with the extended statistics flag -e:

```
# Purpose: Correct systematic error using road
# centerline using raster approach.
# Refer to Ch. 1 Sec. 1.2 for information on where to
# download the data.
import grass.script as grass
grass.run_command( 'r.in.xyz',
   input='road_centerline.txt',
   output='road_centerline', fs=',', x=1, y=2, z=3 )
grass.run_command( 'r.mapcalc',
   expression='temp_NH_2008_1m_error=road_centerline -
   NH_2008_1m' )
# load statistics into a python dictionary with
# parse_command
stats = grass.parse_command( 'r.univar', flags='ge',
   map='temp_NH_2008_1m_error' )
correction = stats['median']
grass.run_command( 'r.mapcalc',
   expression='NH_2008_1m_corrected=NH_2008_1m + ' +
   correction )
```

If the region is large, a vector approach may be more efficient. The benchmarks can be imported as vector points, and a data table can be populated with DEM elevations at benchmark locations using v.what.rast. After updating the database tables, the individual errors can be calculated and the median error can be found by running v.db.univar with the extended statistics -e:

```
# Purpose: Correct systematic error using road
# centerline using vector approach.
import grass.script as grass
grass.run_command( 'v.in.ascii',
   input='road_centerline.txt',
   output='road_centerline', fs=',', x=1, y=2,
   overwrite=True )
```

```
grass.run_command( 'v.db.renamecolumn',
   map='road_centerline', column='dbl_1,x' )
grass.run_command( 'v.db.renamecolumn',
   map='road_centerline', column='dbl_2,y' )
grass.run_command( 'v.db.renamecolumn',
   map='road_centerline', column='dbl_3,z' )

grass.run_command( 'v.db.addcolumn',
   map='road_centerline', layer=1, columns='elev
   DOUBLE PRECISION, error DOUBLE PRECISION' )
grass.run_command( 'v.what.rast',
   map='road_centerline', layer=1,
   raster='NH_2008_1m', column='elev' )
grass.run_command( 'v.db.update',
   map='road_centerline', col='error', qcol='z-elev' )
grass.run_command( 'v.db.select',
   map='road_centerline' )
stats = grass.parse_command( 'v.db.univar',
   flags='ge', table='road_centerline', column='error'
   )
correction = stats['median']
grass.run_command( 'r.mapcalc',
   expression='NH_2008_1m_corrected=NH_2008_1m + ' +
   correction )
```

The approach for correcting systematic error using road lidar data is analogous to using geodetic benchmarks, with the additional step of a centerline extraction. This can be achieved using a least cost path approach where the cost is a function of distance to the sides of the road.

```
import grass.script as grass
# Purpose: Correct systematic error using road surface
# lidar.
# Refer to Ch. 1 Sec. 1.2 for information on where to
# download the data.
grass.run_command( 'r.in.ascii',
   input='DARE_BE94zm3_01m_rstdm.txt',
   output='DARE_BE94zm3_01m_rstdm' )
grass.run_command( 'g.region', flags='pg',
   rast='DARE_BE94zm3_01m_rstdm' )
region = grass.region()
res = region['nsres']
# Generate start raster (side of road).
grass.run_command( 'r.mapcalc',
  expression='DARE_BE94zm3_01m_rstdm_inv=if(isnull
  (DARE_BE94zm3_01m_rstdm),
```

```
  1, null())' )
grass.run_command( 'r.buffer',
   input='DARE_BE94zm3_01m_rstdm_inv',
   output='start_rast', distance=res )
grass.run_command( 'g.remove',
   rast='DARE_BE94zm3_01m_rstdm_inv' )
# Generate a map equal to resolution on the road
# and a map equal to distance from the side of road.
grass.run_command( 'r.mapcalc',
   expression='temp=if(isnull(DARE_BE94zm3_01m_rstdm),
 null(),' + res + ')' )
grass.run_command( 'r.cost', flags='k', input='temp',
   output='cost', start_rast='start_rast' )
# Extract centerline connecting two points
# that were digitized at opposite ends of the road.
pt1='913795,250598'
pt2='913992,250202'
grass.run_command( 'r.mapcalc',
   expression='cost=exp(-5*cost)', overwrite=True )
grass.run_command( 'r.cost', flags='k', input='cost',
   output='ccost', start_coordinates=pt1,
   stop_coordinates=pt2, overwrite=True )
grass.run_command( 'r.drain', flags='n',
   input='ccost', output='NC12_centerline',
   voutput='NC12_centerline', coordinate=pt2 )
```

Once the centerline is extracted, the error correction can be computed using r.mapcalc and r.univar as before:

```
# Purpose: Correct systematic error using
# lidar-extracted centerline.
import grass.script as grass
grass.run_command( 'r.mapcalc',
   expression='NH_2008_1m_error=if(isnull(NC12_
   centerline), null(),
   DARE_BE94zm3_01m_rstdm-NH_2008_1m)' )
stats = grass.parse_command( 'r.univar', flags='ge',
   map='NH_2008_1m_error' )
correction = stats['median']
grass.run_command( 'r.mapcalc',
   expression='NH_2008_1m_corrected=NH_2008_1m + ' +
   correction )
```

References

Abramowitz, M. and Stegun, I. (1965). *Handbook of mathematical functions: with formulas, graphs, and mathematical tables*, volume 55. Dover publications.

Brovelli, M. A., Cannata, M., and Longoni, U. M. (2004). LIDAR data filtering and DTM interpolation within GRASS. *Transactions in GIS*, 8(2):155–174.

Hofierka, J., Parajka, J., Mitasova, H., and Mitas, L. (2002). Multivariate interpolation of precipitation using regularized spline with tension. *Transactions in GIS*, 6(2):135–150.

Mitas, L. and Mitasova, H. (1999). Spatial interpolation. *Geographical Information Systems: Principles, Techniques, Management and Applications, Wiley*, 481.

Mitášová, H. and Mitáš, L. (1993). Interpolation by regularized spline with tension: I. Theory and implementation. *Mathematical geology*, 25(6):641–655.

Mitasova, H., Mitas, L., and Harmon, R. (2005). Simultaneous spline approximation and topographic analysis for lidar elevation data in open-source GIS. *IEEE Geoscience and Remote Sensing Letters*, 2:375–379. DOI: 10.1109/LGRS.2005.848533.

Mitasova, H., Overton, M., Recalde, J., Bernstein, D., and Freeman, C. (2009). Raster-based analysis of coastal terrain dynamics from multitemporal lidar data. *Journal of Coastal Research*, 25:207–215. DOI: 10.2112/07-0976.1.

Neteler, M. and Mitasova, H. (2008). *Open source GIS: a GRASS GIS approach*. New York: Springer, third edition.

Chapter 3
Raster-Based Analysis

Raster-based analysis on two or more DEMs can provide information about change patterns and trends. A common approach to mapping elevation change between two surveys is DEM differencing, performed by map algebra within GIS (r.mapcalc in GRASS). For a larger number of elevation data snapshots, per cell statistics can be applied to the raster DEMs to derive summary maps, which reveal the spatial patterns of stable and dynamic sites, the time periods when sites reach their highest or lowest elevations, and the trends in elevation change.

3.1 Core and Envelope, Dynamic Layer

We can characterize terrain evolution over a given time period by a series of raster DEMs $z(i, j, t_k)$ derived from lidar surveys acquired at time snapshots $t_k, k = 1, \ldots, n$, as discussed in Chap. 2 and Mitasova et al. (2009). We define the *core surface* as the minimum elevation and the *envelope surface* as the maximum elevation measured at each cell over the given time period (t_1, t_n):

$$z_{core}(i, j) = \min_k z(i, j, t_k) \quad k = 1, \ldots, n \tag{3.1}$$

$$z_{env}(i, j) = \max_k z(i, j, t_k) \quad k = 1, \ldots, n \tag{3.2}$$

The space bounded by the core and envelope surfaces defines a *dynamic layer*. For a sandy coastal environment, typical for barrier islands, the core surface represents a boundary between a dynamic layer and a stable sand volume that has not moved during the entire study period. The envelope surface represents the outer boundary of the dynamic layer within which the terrain evolved during the given time period (t_1, t_n).

© The Author(s) 2014
E. Hardin et al., *GIS-based Analysis of Coastal Lidar Time-Series*, SpringerBriefs in Computer Science, DOI 10.1007/978-1-4939-1835-5_3

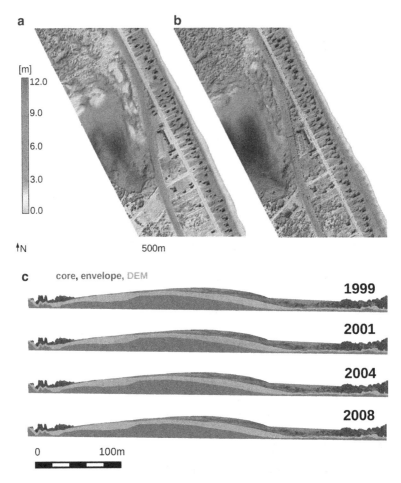

Fig. 3.1 (**a**) Core and (**b**) envelope surfaces derived from time series of lidar data (**c**) cross section of the core, envelope, and terrain surface at multiple time snap shots

We run the r.series command with the method parameter set to minimum compute the core surface and maximum to compute the envelope surface (Fig. 3.1). 'No-data' values are propagated using the -n flag when computing the core surface because 'no-data' values indicate areas that were subaqueous in the time series:

```
# Purpose: Compute core and envelope surfaces.
import grass.script as grass
grass.run_command( 'g.region', rast='NH_1999_1m' )
mlist = grass.read_command( 'g.mlist', type='rast',
    pattern='NH_*_1m', mapset='.', fs=',' )
# Compute the core surface.
```

```
grass.run_command( 'r.series', flags='n', input=mlist,
    output='NH_core', method='minimum' )
grass.run_command( 'r.colors', map='NH_core',
    raster='NH_1999_1m' )
# Compute the envelope surface.
grass.run_command( 'r.series', input=mlist,
    output='NH_env', method='maximum' )
grass.run_command( 'r.colors', map='NH_env',
    raster='NH_1999_1m' )
```

Core and envelope surfaces provide the basis for several quantitative measures of coastal dynamics such as shoreline migration range and relative volume change. We will cover these measures later in this book.

3.2 Time-of-Minimum and Time-of-Maximum

A raster map representing the time associated with the core surface can be used to identify locations and time when the land surface was at its minimum within the given study period. Similarly, raster map showing the time associated with the envelope surface identifies the time when the land surface was at its maximum. Raster maps representing *time-of-minimum* and *time-of-maximum* can be computed as:

$$t_{max}(i, j) = t_l, \quad \text{where} \quad z(i, j, t_l) = z_{env}(i, j) \tag{3.3}$$

$$t_{min}(i, j) = t_p, \quad \text{where} \quad z(i, j, t_p) = z_{core}(i, j) \tag{3.4}$$

where the indices l and p represent the values in the time map and t_l or t_p, the date at which this value occurred, is stored as an attribute (label).

We compute the time-of-minimum and time-of-maximum maps by running r.series with the method parameter set to min_raster and max_raster, respectively (Fig. 3.2). We then use the r.category command to assign the labels representing the dates associated with the index of the raster maps in the time series.

```
# Purpose: Compute time-of-minimum and time-of-maximum
# maps, and apply categories.
import tempfile
import os
fname = tempfile.mkstemp()[1]
f = open( fname, 'w' )
rules = """0:1999
1:2001
2:2004
3:2005
```

```
4:2007
5:2008"""
f.write( rules )
f.close()
grass.run_command( 'r.series', flags='n', input=mlist,
    output='NH_min_time', method='min_raster' )
grass.run_command( 'r.series', input=mlist,
    output='NH_max_time', method='max_raster' )
grass.run_command( 'r.category', map='NH_min_time',
    rules=fname )
grass.run_command( 'r.category', map='NH_max_time',
    rules=fname )
os.remove( fname )
```

In our sample data, the time of maximum map shows clearly that the elevation maximum has moved south over time (Fig. 3.2b).

3.3 Per-Cell Univariate Statistics

In addition to core, envelope, and time-of-minimum and time-of-maximum maps, other univariate statistics with physical interpretations can be calculated. The standard deviation map can be used to identify the stable and dynamic areas in terms of elevation change. It is computed as:

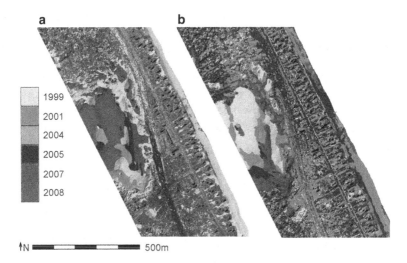

Fig. 3.2 Surfaces representing the time of (**a**) minimum and (**b**) maximum elevation observed in the time series of lidar data

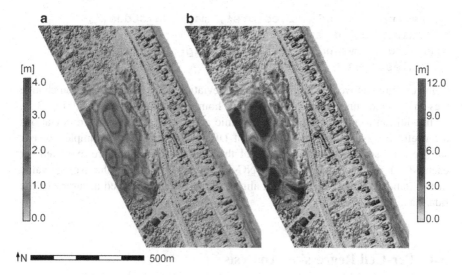

Fig. 3.3 Surfaces representing (**a**) the standard deviation and (**b**) range of terrain elevation observed in the time series of lidar data

$$z_\sigma(i, j) = \sqrt{\frac{1}{n} \sum_{k=1}^{n} [z(i, j, t_k) - z_\mu(i, j)]} \qquad (3.5)$$

where

$$z_\mu(i, j) = \frac{1}{n} \sum_{k=1}^{n} z(i, j, t_k) \qquad (3.6)$$

where $z_\sigma(i, j)$ is the standard deviation in the raster cell (i, j) and $z_\mu(i, j)$ is the mean elevation value observed in this cell over time. The range map, computed as the difference between the maximum and minimum elevation recorded during the study period represents the magnitude of elevation change, or the thickness of the dynamic layer at each raster cell.

We compute the standard deviation and range maps by running `r.series` with the `method` parameter set to `stddev` and `range`, respectively (Fig. 3.3):

```
# Purpose: Compute standard deviation and range maps.
grass.run_command( 'r.series', input=mlist,
    output='NH_stddev', method='stddev' )
grass.run_command( 'r.series', input=mlist,
    output='NH_range', method='range')
# Specify color maps in the current directory (or use a
# full file path name).
```

```
grass.run_command( 'r.colors', map='NH_stddev',
   rules='color_stddev.txt' )
grass.run_command( 'r.colors', map='NH_range',
   rules='color_range.txt' )
```

In the example above, the highest standard deviation and range of elevation change was on the sand dune and at locations where homes were built or removed (Fig. 3.3).

In addition to mapping the most dynamic and stable areas, standard deviations are useful for evaluating the accuracy of DEM time series. For example, for our sample time series, when the values of the standard deviations are extracted for each NC DOT benchmark along the NC-12 highway centerline, the average value of the standard deviation is 0.14 m, almost equal to the published accuracy of the lidar data.

3.4   Per-Cell Regression Analysis

Continuous terrain evolution is characterized by gradual changes in the elevation surface over time; For example, such changes may represent sand dune migration due to wind transport or vegetation growth. To detect discrete changes, we extracted elevation changes that exceeded a set threshold; Whereas, to quantify continuous evolution, we map spatial patterns and the rate of elevation change over time.

The spatial distribution of the linear rate of change can be estimated by computing linear regression for each raster cell (i, j) in the time series of n raster elevation maps. The result of the regression can be represented by three raster maps: (a) slope of the regression line $r_s(i, j)$; (b) offset $r_o(i, j)$; and (c) coefficient of determination $r_c(i, j)$.

$$z_r(i, j) = r_o(i, j) + r_s(i, j)z(i, j) \tag{3.7}$$

For active bare wind blown dunes, both elevation growth and loss can be continuous. We can map the dune erosion and dune growth areas as raster cells (i_e, j_e) and (i_d, j_d), respectively, that fulfill the following conditions

$$r_s(i_e, j_e) < \varepsilon_e \cap r_c(i_e, j_e) > r_{cmin} \tag{3.8}$$

$$r_s(i_d, j_d) > \varepsilon_d \cap r_c(i_d, j_d) > r_{cmin} \tag{3.9}$$

where ε_e and ε_d are threshold negative and positive regression slopes indicating dune erosion and growth, respectively, and r_{cmin} is a threshold value for the coefficient of determination for which the relationship can be considered linear. Areas with $r_c(i, j)$ less than r_{cmin} do not a have clear linear trend of growth or decline. These represent areas where growth has switched to decline as the dune has migrated. Similar analysis can be applied to forest canopy surfaces to estimate the rate of forest growth and identify areas with forest decline.

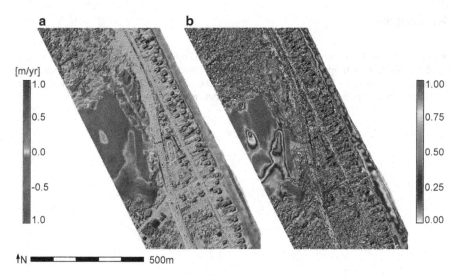

Fig. 3.4 Surfaces derived from the time series of lidar: (**a**) the rate of elevation change and (**b**) coefficient of determination for which the rate of change can be considered linear

The r.series command assumes that input maps are spaced at even time intervals. If input maps are not at even time intervals, maps containing 'no-data' values can be inserted so that maps are spaced approximately as they should be. We can then compute the linear regression maps shown in Fig. 3.4 as follows:

```
# Purpose: Compute null maps for use when elevation
# snapshots are not evenly spaced in time.
import grass.script as grass
grass.run_command( 'r.mapcalc',
   expression='Null=null()' )
mlist =
   'NH_1999_1m,Null,NH_2001_1m,Null,Null,NH_2004_1m,
    NH_2005_1m,Null,NH_2007_1m,NH_2008_1m'
grass.run_command( 'r.series', input=mlist,
   output='NH_r_s,NH_r_o,NH_r_c',
   method='slope,offset,detcoeff' )
grass.run_command( 'r.colors', map='NH_r_s',
   rules='color_regrslope.txt' )
grass.run_command( 'r.colors', map='NH_r_c',
   rules='regrcoefdet.txt' )
```

An alternate approach, registers the DEMs in a temporal database using the commands t.create and t.register. Then the regression can be performed by the t.rast.series command which supports variable time interval.

References

Mitasova, H., Hardin, E., Overton, M., and Harmon, R. (2009). New spatial measures of terrain dynamics derived from time series of lidar data. *Proc. 17th Int. Conf. on Geoinformatics, Fairfax, VA*. DOI: 10.1109/GEOINFORMATICS.2009.5293539. Associated animation: http://skagit.meas.ncsu.edu/$\sim$helena/gmslab/papers/Core_Envelope_anim.ppt.

Chapter 4
Feature Extraction and Feature Change Metrics

Coastal change has been historically measured by metrics derived for specific coastal linear features such as shorelines. Linear features are also important for measuring sand dune migration based on the location of dune crests and slip faces and for prediction of coastal vulnerability. In this chapter we present methods for extracting shorelines, dune ridges, dune crests and building footprints from DEMs. Then we measure the change of these features and use them to map vulnerability to storms.

4.1 Shorelines and Shoreline Migration Range

A shoreline is the interface between the land and the ocean. The location of the shoreline is fundamentally important to various aspects of coastal science and engineering. However, the cross-shore and vertical location of the shoreline continuously changes both temporally and spatially. Thus, differing definitions of the shoreline exist for various applications.

Shoreline location is often approximated by a constant elevation contour drawn in a tidal datum. The specific tidal datum and contour elevation may depend on the application. With our set of sample DEMs registered relative to the mean high water datum, we approximate the shoreline location with the $z = 0.8$ m elevation contour, extracted with the r.contour command. We use an elevation of 0.8 m (which was selected by Burroughs and Tebbens (2008)) because elevation data seaward of the 0.0 m shoreline proxy have already been set to 'no-data' values in the sample DEMs. To eliminate small 0.8 m contours on the beach representing small depressions, we use the parameter cut which removes all contours with a small point count (below the given threshold), leaving only a single shoreline proxy.

© The Author(s) 2014
E. Hardin et al., *GIS-based Analysis of Coastal Lidar Time-Series*, SpringerBriefs in Computer Science, DOI 10.1007/978-1-4939-1835-5_4

```
# Purpose: Use r.contour to extract contour.
# 1 is the shoreline proxy elevation
# c is the threshold to exclude small anomalies
l = 0.8
c = 400
dates = [19961016, 19971002, 19980907, 19990909,
    19990918, 2001, 20040925, 20051126, 20080327]
for date in dates:
    rast = 'NH_'+str(date)+'_1m'
    grass.run_commands( 'r.contour', input=rast,
        output='NH_'+str(date)+'_08m', levels=l,
        cut=c)
```

The $z = 0.8$ m elevation contour extracted from the 2008 DEM as an approximation of the shoreline is shown in Fig. 4.4. Extracted contours can be smoothed using the v.generalize command. A number of smoothing approaches are available, which are set via the method parameter. A sliding average approach can be employed by running:

```
# Purpose: Smooth extracted shoreline.
v.generalize input=NH_2008_08m output=NH_2008_08m_sm \
    method=sliding_averaging threshold=0 look_ahead=51
```

Shoreline contours extracted from the core and envelope surfaces define a *shoreline migration range* within which the shoreline evolved during the given time interval (t_1, t_n). The short-term range of shoreline migration is the distance between the most ocean-ward (maximum) and most landward (minimum) locations of the shoreline within the study period. The shoreline might have migrated only in one direction (landward in case of systematic erosion or ocean-ward in case of systematic accretion) or back and forth (e.g., erosion due to hurricane followed by recovery, or due to the erosion-nourishment-erosion cycle). The shoreline migration range (also referred to as the shoreline band) is shown in Fig. 4.1.

Shoreline-based metrics such as erosion/accretion rate and migration range do not measure the volume of sediment moved or the location of the displaced volume. For example, while many storm events erode the beach, the sediment is carried into the nearshore and beach recovery can occur after the storm as the waves build the beach. On the other hand, during a washover event, both sediment from the beach and the dune may be carried landward and taken out of the beach and dune system. The short-term range of shoreline migration quantifies the variability of shoreline position, but does not provide long-term trend information (Mitasova et al. 2012).

4.2   Foredune Features

Foredunes are linear dunes parallel to the shoreline rising on the in-land side of the beach. These dunes provide critical protection for homes and roads during storms. Properties of the foredunes, such as the ridge height and position of its toe

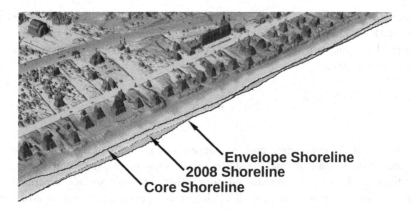

Fig. 4.1 The shoreline migration range in the town of Nags Head

relative to the beach are fundamental parameters for estimating coastal vulnerability. In this section, we present innovate techniques for extracting foredune features using GIS-based tools.

4.2.1 Foredune Ridge Line

Traditionally, foredune ridges have been mapped using cross-shore transects or by digitizing based on the data derived from DEMs and imagery. Here we present an alternative, more automated method based on the least cost path tracing (Hardin et al. 2012).

To use the least cost path to derive the fore dune ridge line, a suitable quantitative definition of the dune ridge is needed. In this way, a cost surface can be generated and the least cost path can be found. There are a number of conditions that the cost function must satisfy. First, the cost function must be an inverse function of elevation because the dune ridge is at a local elevation maximum and the paths that pass below the dune ridge should have high cumulative costs. Second, the least cost path is found based on the cost of passing a grid cell and the traveled distance. To account for the complexity of dune ridges, the cost of a slightly shorter, lower elevation path should be greater than a slightly longer, higher elevation path. Third, the cost function should be continuous everywhere, unlike z^{-1} and other power functions, which can satisfy the above two conditions but are discontinuous at $z = 0$. Finally, the cost function should be positive everywhere. This means that the cost should approach zero as elevation increases as opposed to approaching negative infinity. The cost function that was found to fulfill the above stated conditions was defined as

$$J_{ij} = e^{-\alpha z_{ij}} \tag{4.1}$$

where J_{ij} is the dimensionless cost of traversing the raster cell (i, j), z_{ij} (m) is the DEM elevation, and α (m^{-1}) is a tunable parameter. The dune ridge is then extracted by generating the cumulative cost surface and finding the path with the least cumulative cost:

$$J_{tot} = \min_n \sum_n^N J_{i_n, j_n} \qquad (4.2)$$

where J_{tot} is the cumulative cost, n indexes over the cells in the path, N is the number of cells in a path, and the starting and ending points, (i_0, j_0) and (i_N, j_N), are fixed at the ends of the studied dune ridge. The cost surface described by Eq. (4.1) can be generated in a GIS using map algebra, and the least cost path described by Eq. (4.2) can be calculated using standard cumulative cost surface generation and least cost path routing tools.

```
# Purpose: Extract a dune ridge as a least cost path.
# Specify two points that were manually selected at
# opposite ends of dune ridge.
# Compute cost surface.
r.mapcalc expression='cost=exp(-2*NH_2008_1m)' --o
# Compute a cumulative cost surface.
r.cost -k input=cost output=cumulative_cost \
    start_coordinates='913859,250658' \
    stop_coordinates='914305,249739' --o
# Calculate the least coast path.
r.drain input=cumulative_cost \
    output=NH_2008_duneRidge \
    vector_output=NH_2008_duneRidge \
    start_coordinates='914305,249739' --o
# Extract dune ridge.
r.mapcalc \
  expression='NH_2008_duneRidge=float\
  (NH_2008_duneRidge)*NH_2008_1m' --o
```

The dune ridge extracted from the 2008 DEM is shown in Fig. 4.4. In this example, which performs ridge-line extraction on a DEM with homes, the algorithm runs the line across two homes which are very close to the actual dune ridge. Later in this chapter, we will show how to extract building footprints and create a least cost path which avoids the homes.

The complexity of the least cost path line geometry can be adjusted using the tunable parameter α which has been set to a value of 2 in our example, based on empirical observations. High values of α yield a more complex, detailed ridge line shape; Whereas, lower values of α lead to a straighter, simplified line (Fig. 4.2). Values for α can be optimized for a given application using a representative sample of the dune ridge where a highly accurate location of the dune ridge is available. The optimized α will also depend on the scale of the process being modeled.

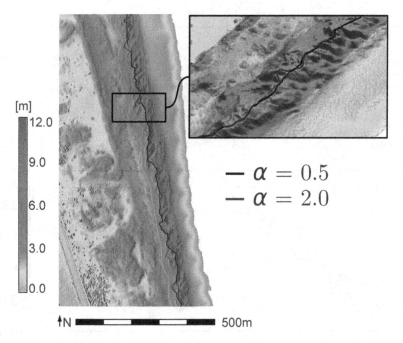

Fig. 4.2 Ridge lines extracted with different values of parameter α

4.2.2 Foredune Toe Line

The foredune toe is qualitatively defined as the location where the beach meets the coastal foredune. It is often identified along a cross-shore beach profile as the location with the greatest change in slope (profile curvature maximum) or where shallow sloping terrain on the beach meets steep foredune terrain (Burroughs and Tebbens 2008; Elko et al. 2002; Stockdon et al. 2007; Stockdon and Thompson 2007a,b). Estimating the curvature requires the computation of second-order derivatives of an intrinsically noisy surface. This makes the curvature-based dune toe extraction methods highly dependent on resolution and smoothing parameters.

Alternatively, the dune toe can be identified as the location with the largest distance between the given elevation profile and a line connecting the dune ridge and shoreline (Mitasova et al. 2011). A continuous dune toe line can be extracted by implementing this conceptualization in two dimensions, where the beach profile becomes the terrain surface and the line connecting the dune ridge and shoreline becomes a tightly stretched elastic sheet with boundary conditions at the dune ridge and shoreline (Fig. 4.3). Further, instead of approximating the dune toe as the point where the elevation profile is most deviated from a line connecting the dune ridge and shoreline, the path of greatest deviation between the terrain surface and the elastic sheet approximates the continuous dune toe line (Fig. 4.3).

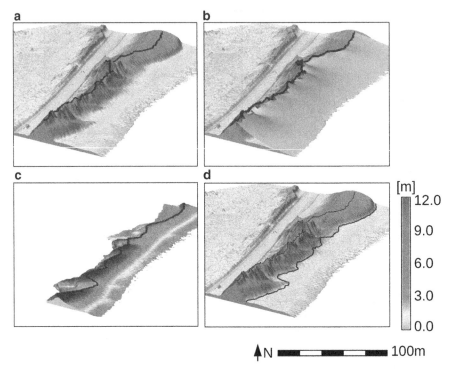

Fig. 4.3 Illustration of the extraction of a dune toe in which (**a**) the dune ridge is extracted, (**b**) the elevation of a zero-energy elastic membrane connecting the dune ridge and the shoreline is computed, (**c**) the sheet and the terrain are differenced and the path of greatest deviation is computed, and finally, (**d**) the dune toe (approximated by the path of greatest deviation) is overlaid on the terrain surface

The elastic sheet can be modeled as an array of springs with nodes located at each raster cell (Hardin et al. 2012), which is a crude model for a deformable cloth that is well established in computer graphics literature (Breen et al. 1992; Provot 1995). The nodes at the ends of the sheet are fixed to the shoreline and the dune ridge. Only a spring force acts on each node, which depends on its elevation relative to the elevation of its neighbors,

$$F_{i,j} = (z_{i,j} - z_{i+1,j}) + (z_{i,j} - z_{i-1,j}) + (z_{i,j} - z_{i,j+1}) + (z_{i,j} - z_{i,j-1}) \quad (4.3)$$

$$= 4z_{i,j} - z_{i+1,j} - z_{i-1,j} - z_{i,j+1} - z_{i,j-1} \quad (4.4)$$

The static equilibrium position of the sheet can efficiently be computed by solving Eq. (4.3) as a linear system of N equations where N is the combined number of nodes that either constitute the sheet or represent boundary conditions. This approach requires software functionality not typically included in standard GIS packages. Alternatively, we can compute Eq. (4.3) using an iterative, relaxation approach:

$$z_{i,j}^* = \frac{1}{4}(z_{i+1,j} + z_{i-1,j} + z_{i,j+1} + z_{i,j-1}) \qquad (4.5)$$

where with each iteration the elevation of every node is reassigned the mean elevation of its four nearest neighbors during the previous iteration. This is achieved by repeatedly applying a Laplace filter (or smoothing filter). We apply a Laplace filter in GRASS using the r.neighbors command where the neighborhood operation is set to average (method=average) and the neighborhood is limited to the four nearest neighbors by setting size=3 and using the -c flag to indicate a circular neighborhood.

Once the elevation of the elastic sheet is solved, we difference the elevations of the sheet and the terrain surface to produce a raster map that represents the deviation between the two surfaces, $D_{i,j}$ (Fig. 4.3). Finally, we derive the cost surface using Eq. (4.1) where $z_{i,j}$ has been replaced by the difference between the DEM and the sheet, $D_{i,j}$, and we extract the dune toe line as the least cost path using Eq. (4.2). The extracted dune toe line is shown in Fig. 4.4. The following GRASS code implements the dune toe extraction process:

```
# Purpose: Compute dune toe line with elastic sheet
# method.
#
# Reduce computational region to area containing
# boundary
# conditions for efficiency.
import grass.script as grass
grass.run_command( 'v.patch',
    input='NH_2008_duneRidge,NH_2008_08m',
    output='sheet_BC' )
grass.run_command( 'g.region', vect='sheet_BC' )
grass.run_command( 'v.to.rast', input='NH_2008_08m',
    output='NH_2008_08m', use='val', value='0.8' )
grass.run_command( 'r.patch',
    input='NH_2008_duneRidge,NH_2008_08m',
    output='sheet_BC', overwrite=True )
grass.run_command( 'g.copy', rast='NH_2008_1m,sheet' )
iterations = 3000
for i in range(iterations):
    print i
    grass.run_command( 'r.neighbors', flags='c',
        input='sheet', output='sheet', method='average',
        size=3, overwrite=True )
    grass.run_command( 'r.patch',
        input='sheet_BC,sheet', output='sheet',
        overwrite=True )
```

```
grass.run_command( 'r.colors', map='sheet',
   rast='NH_2008_1m' )

# Make small null buffer around dune ridge and
# shoreline to keep extracted toe between them.
grass.run_command( 'r.buffer', input='sheet_BC',
   output='sheet_BC_buff', dist=1 )
grass.run_command( 'r.mapcalc',
   expression='deviation_map=if(isnull(sheet_BC_buff),
   sheet-NH_2008_1m, null())' )
# Again, use two manually selected points.
pt1 = '913878,250654'
pt2 = '914317,249759'
# Extract dune toe.
grass.run_command( 'r.mapcalc',
   expression='cost=exp(-5*deviation_map)' )
grass.run_command( 'r.cost', flags='k', input='cost',
   output='cumulative_cost', start_coordinates=pt1,
   stop_coordinates=pt2 )
grass.run_command( 'r.drain', input='cumulative_cost',
   output='NH_2008_duneToe',
   voutput='NH_2008_duneToe', start_coordinates=pt2 )
grass.run_command( 'r.mapcalc',
   expression='NH_2008_duneToe=float(NH_2008_duneToe)
   *NH_2008_1m' )
```

4.3   Crescentic and Parabolic Dune Features

Sand dune fields with parabolic or crescentic dunes form in coastal areas with a large supply of sand and steady, strong winds. These dunes, also referred to as backdunes, often migrate and change their shape creating a highly dynamic landscape. To quantify the dune migration and transformation we extract dune features, such as peaks, slip faces, active crests, and windward side ridges. Slip faces and dune crests can be extracted using thresholds in slope and profile curvature.

We can compute these parameters simultaneously by interpolating DEMs using the first- and second-order partial derivatives of the RST function and principles of differential geometry (Mitasova and Hofierka 1993). The same general equations apply for estimating these parameters from raster-based DEMs. Assuming that the elevation surface is approximated by a bivariate function $z = f(x, y)$ (which can be the RST function or a polynomial), we first we introduce the following simplifying notations:

$$f_x = \frac{\partial z}{\partial x}, \qquad f_y = \frac{\partial z}{\partial y}, \qquad f_{xx} = \frac{\partial^2 z}{\partial x^2}, \qquad f_{yy} = \frac{\partial^2 z}{\partial y^2}, \qquad f_{xy} = \frac{\partial^2 z}{\partial x \partial y}$$

$$(4.6)$$

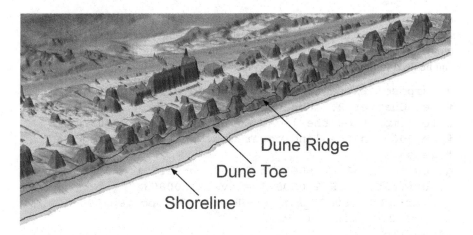

Fig. 4.4 DEM overlaid with the extracted shoreline (0.8 m elevation contour), dune toe, and dune ridge. The dune ridge extraction approach erroneously extracts structures when structures are on or immediately adjacent to the dune ridge. Using bare ground DEM or creating a cost surface with high cost for pixels with buildings will correct this problem as we show later in this chapter

and

$$p = f_x^2 + f_y^2 , \quad q = p + 1. \qquad (4.7)$$

Then the steepest slope angle γ in degrees or percent is computed from gradient $\nabla f = (f_x, f_y)$ as follows

$$\gamma = \arctan\sqrt{p} \qquad \gamma[\%] = 100.\sqrt{p} \qquad (4.8)$$

and the equation for the profile curvature $\kappa_s (m^{-1})$ is

$$\kappa_s = \frac{f_{xx} f_x^2 + 2 f_{xy} f_x f_y + f_{yy} f_y^2}{p \sqrt{q^3}}. \qquad (4.9)$$

Previous application of this approach to lidar data demonstrated that suitable selection of the RST parameters (tension and smoothing) is essential for deriving the topographic parameters at the level of detail matching the size of the dune features (Mitasova et al. 2004). It is also important to note that the analysis should be applied to a bare earth DEM. If only the first return DEM is available, the vegetated areas should be masked out.

In our example, we will extract the dune crests and slip faces from the first return lidar point cloud which represents the Jockey's Ridge dune field in the year 2008. First, we interpolate the DEM with simultaneous computation of slope and profile curvature. We use 3D visualization and raster map query to identify the threshold values of profile curvature and slope associated with the dune crests and slip faces. Then, we use map algebra to mask out the vegetated areas based on the land cover map derived by Weaver et al. (2010) and to extract dune crests and slip faces into separate raster map layers. The threshold of profile curvature $\kappa_s > 0.08\,\mathrm{m}^{-1}$ (convexity) proved to be a good indicator of dune crests (Fig. 4.5a). Slopes $\gamma > 25°$ can be used to identify the slip faces (Fig. 4.5b).

```
# Purpose: Extract dune crests and slip faces.
# see Chapter 2, subsection 2.2.3
# for import of the lidar point cloud JR_20080327
# lu_2009 is the land cover map, where the category 1
# is sand
g.region res=1 n=250690 s=249502 w=912791 e=913931 -p
v.surf.rst -t JR_20080327 elev=JR_20080327 \
    slo=JR_20080327_slp pc=JR_20080327_pc ten=500 \
    smo=2. npmin=150 dmin=0.9
r.mapcalc \
 expression='JR_20080327_crest=if(JR_20080327_pc>0.04\
    && lu_2009 == 1,1,null())'
r.mapcalc \
    expression='JR_20080327_slip=if(JR_20080327_slp>25 \
    && lu_2009 == 1,2,null())'
```

In addition to the dune crests and slip faces, backslope dune ridges and evolution of slope values along these ridges are important indicators of dune landform stages in relation to sand supply. In general, mountain ridges can be identified using plan or tangential curvature (Mitasova and Hofierka 1993); However, this approach is not very suitable for dunes with smooth windward sides and low values of curvature. Therefore, an alternative approach based on the density of slope lines generated uphill from each grid cell can be used (Mitasova et al. 1996). The slope lines follow the direction of surface gradient and are perpendicular to contours. This approach is an inverse version of the flow accumulation algorithm commonly used for extraction of streams (e.g., Tarboton 1997). To extract ridges, slope lines are computed using a D-infinite (vector-grid) algorithm (Mitasova et al. 1996). This approach avoids the artificial patterns produced by the standard D-8 methods on smooth surfaces typical for dunes. Dune ridges are then extracted as grid cells with the number of slope lines passing through them (slope line accumulation) greater than a given threshold. In the following example, the threshold is 600: (Fig. 4.6).

```
# Purpose: Compute upslope flowline density raster
# map and
# vector representation of flowlines for each
# 50th cell.
r.flow -u JR_20080327_05m dsout=JR_20080327_05m_upfl
# flout=JR_20080327_05m_upfline skip=50
# Extract raster cells with flowline density > 600
# and assign them the slope values.
r.mapcalc\
```

```
expression='JR_20080327_05m_ridges=if(/
  JR_20080327_05m_upfl > 600, slope, null())'
# Compute the univariate statistics for the slope
# along the ridge line.
r.univar JR_20080327_05m_ridges
```

We can apply this approach to the entire time series of DEMs for Jockey's Ridge and analyze the trend in slope values along the ridges. Decline in the average slope along the ridge would be an indication of dune stabilization.

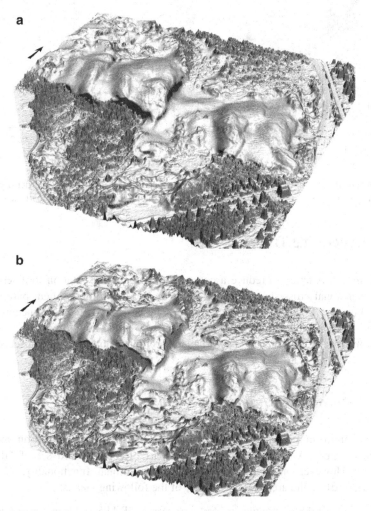

Fig. 4.5 DEM overlaid with the map of (**a**) slip faces and (**b**) dune crests

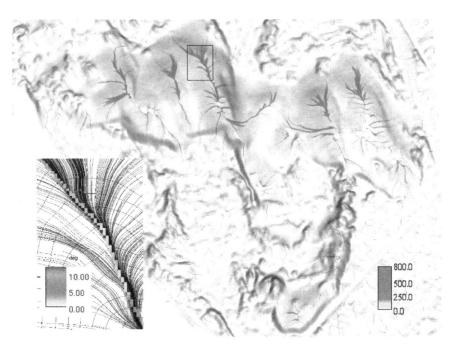

Fig. 4.6 DEM (year 2001, bare earth) overlaid with the map of uphill slope-line density color map. *Inset* shows the slope-lines and the values of slope along the extracted section of the ridge

4.4   Generating Transects

It is common practice to reduce a coastal landscape representation to a series of two-dimensional cross-shore profiles. From the set of profiles, salient parameters (e.g., shoreline location and beach slope) can be measured and conceptual or physics-based models can be computed. Here we outline how to generate a set of uniformly spaced transects and how to adaptively place non-uniform transects.

4.4.1   Transects at Uniform Locations

A set of uniform profiles can be drawn in GRASS using the add-on module `v.transects`. Add-on modules are not part of the standard set of GRASS modules; However, add-on modules include additional functionality, such as `v.transects`, and are available by any of the following means:

- Use the GRASS module `g.extension`. GRASS add-ons are GRASS extensions written by members of the GRASS community.
- Download from the GRASS add-ons page: http://grass.osgeo.org/download/addons/

- Download from (or upload to) the source code repository using the SVN client software:

```
svn co http://svn.osgeo.org/grass/grass-addons
    /grass7
```

The module v.transects draws a set of transects that are orthogonal to a baseline (e.g., shoreline or other contour). The transects are uniformly spaced according to the transect_spacing parameter, and are offset from the baseline according to the dleft and dright parameters. For example, a set of transects spaced at 50 m intervals, as shown in Fig. 4.7, can be generated along the 0.8 m contour by running the following GRASS commands:

```
# Purpose: Generate a set of cross-shore transects.
g.region rast=NH_2008_1m
r.contour input=NH_2008_1m output=NH_2008_08m \
    level=0.8 cut=400
v.transects.py input=NH_2008_08m \
    output=NH_2008_transects transect_spacing=50 \
    dleft=50 dright=150 --o
```

4.4.2 Transects at Optimized Locations

The choice of profile spacing is a balance between resources and adequate representation of the terrain. However, profiles can be irregularly spaced in an optimized way to best represent the terrain using an approach based on line simplification.

Measurements of topographic parameter values along a series of transects can be thought of as a polyline in the parameter space of topographic parameters. The polyline representation is specified by a sequence of n vertices $(V_1, V_2, \ldots, V_n)$ and $n - 1$ line segments $(\overline{V_1 V_2}, \overline{V_2 V_3}, \ldots, \overline{V_{n-1} V_n})$. The vertices correspond to transects and have k coordinates, which correspond to alongshore location and $k - 1$ topographic parameter values.

Profiles can be irregularly spaced by first constructing a polyline representation of the topography at the maximum alongshore resolution. The number of vertices in the polyline representation (which correspond to profiles) can be considerably reduced while still representing the terrain well using the Douglas and Peucker (1973) line simplification algorithm. The Douglas and Peuker algorithm approximates a polyline, P, with a polyline, P', where the vertices that specify P' are a subset of the vertices that specify P (Ebisch 2002; Hershberger and Snoeyink 1992). The vertices that specify P' are determined in a recursive manner until an error threshold, ϵ, is met.

Initially, the simplified polyline is specified only by the endpoints of the unsimplified polyline,

$$P' = \overline{V_1 V_n}. \tag{4.10}$$

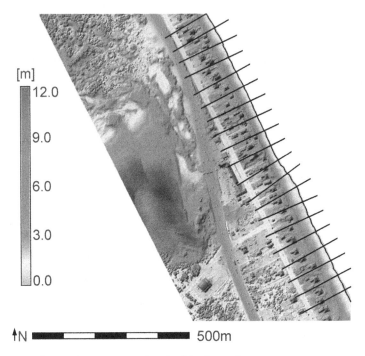

Fig. 4.7 A set of cross-shore transects placed at 50 m intervals

Then, the vertex $V_f \in \{V_1, V_3, \ldots, V_n\}$ that is most separated from P' is identified. If V_f is separated from P' by a distance less than ϵ, then P' is considered to be a good approximation to P, and the simplification of P is complete. Alternatively, if the distance from V_f to P' is greater than or equal to ϵ, then V_f is inserted into P',

$$P' = \overline{V_1 V_f V_n}. \tag{4.11}$$

Following the inclusion of V_f, the procedure is recursively applied to each line segment in P' with the corresponding set of vertices (e.g., $\overline{V_1 V_f}$ with $\{V_1, \ldots, V_f\}$ and $\overline{V_f V_n}$ with $\{V_f, \ldots, V_n\}$) until a good approximation to P is reached. Finally, the profiles that correspond to the remaining vertices are used to model the terrain whereas the others are discarded.

Prior to polyline simplification, parameter space should be normalized so that the simplification is not biased to represent the parameters that typically assume larger values (e.g., beach width compared to dune height). The vertex coordinates in parameter space can be normalized according to:

$$\hat{x}_n^p = \frac{\mathbf{x}_n^p - <\mathbf{x}^p>}{\sigma_x^p} \tag{4.12}$$

where $\hat{x}_n^p$ is the pth component of the normalized coordinate of the nth vertex, p is the normalized component (which corresponds to a particular topographic parameter), $\mathbf{x}_n^p$ is the unnormalized component of the coordinate, $< \mathbf{x}^p >$ and σ_x^p are the mean and standard deviation respectively of the pth component of all the vertices in the polyline.

We generate the polyline representation of an 1D feature by replacing the x,y values with cat,z values. Then we simplify the line using v.generalize by setting the method parameter to douglas_reduction:

```
# Purpose: Optimize transect locations.
import grass.script as grass
import tempfile

grass.run_command( 'g.region', rast='NH_2008_1m' )
# Extract dune ridge.
pt1='913855,250657'
pt2='914305,249740'
grass.run_command('r.mapcalc',
    expression='cost=exp(-5*NH_2008_1m)',
    overwrite=True )
grass.run_command('r.cost', flags='k', input='cost',
    output='cumulative_cost', start_coordinates=pt1,
    stop_coordinates=pt2, overwrite=True )
grass.run_command('r.drain', flags='n',
    input='cumulative_cost',
    output='NH_2008_duneRidge', start_coordinates=pt2,
    overwrite=True )

# Output dune ridge and import as point vector map
fname = tempfile.mkstemp()[1]
grass.run_command( 'r.stats', flags='1gn',
    input='NH_2008_duneRidge,NH_2008_1m', output=fname,
    fs='|' )
grass.run_command( 'v.in.ascii', input=fname,
    output='test', x=1, y=2, cat=3, z=4, fs='|',
    overwrite=True )

# find the number of points
vinfo = grass.parse_command( 'v.info', flags='tg',
    map='test' )
points = vinfo['points']

# Import parameter-space line by changing the category
# values to be the x value and the elevation to be
# the y value
```

```python
# Import as a line by converting the result to
# grass standard vector format
vect = grass.read_command( 'v.out.ascii', input='test'
    )
vect = vect.strip('\n').split('\n')
vect = [ line.split('|') for line in vect ]
vect = [ [int(line[3]),float(line[2])] for line in
    vect ]
vect.sort()
vect = [ ' '.join(map(str,line)) for line in vect ]
vect = '\n'.join( vect )
vect = 'L ' + points + '\n' + vect

fout = open( fname, 'w' )
fout.write( vect )
fout.close()
grass.run_command( 'v.in.ascii', flags='n',
    input=fname, output='test_ps', format='standard',
    overwrite=True )

# Simplify parameter space line.
grass.run_command( 'v.generalize', flags='c',
    input='test_ps', output='test_ps_gen',
    method='douglas_reduction', threshold=0,
    reduction=10, overwrite=True )

grass.run_command( 'v.to.points', flags='v',
    input='test_ps_gen', output='test_ps_gen_pts',
    overwrite=True )

vinfo = grass.parse_command( 'v.info', flags='tg',
    map='test_ps_gen_pts', overwrite=True )
points = vinfo['points']

# retrieve the categories (called "cats") of the nodes
# that remain in the simplified line
cats = grass.read_command( 'v.out.ascii',
    input='test_ps_gen_pts' )
cats = cats.strip('\n').split('\n')
cats = [ int( line.split('|')[0] ) for line in cats ]
cats.sort()
cats = map( str, cats )
cats = ','.join( cats )
```

```
grass.run_command( 'v.out.ascii', input='test',
    output=fname, cats=cats )
grass.run_command( 'v.in.ascii', input=fname,
    output='NH_duneRidge_gen_pts', x=1, y=2, z=3,
    cat=4, fs='|', overwrite=True )
os.remove( fname )
```

Figure 4.8 shows a set of transects placed at optimized locations so that they capture the variations of a dune ridge.

4.5 Measuring Line Feature Change

Feature change is usually measured as a horizontal distance between the location of features at different time snapshots. The distance is measured along regularly spaced transects and the result is often reported as a spatially (and temporally) aggregated value.

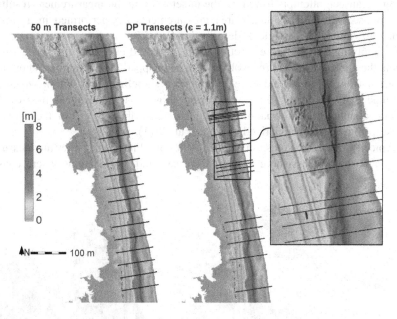

Fig. 4.8 A set of cross-shore transects placed at regular and optimized locations

4.5.1 Shoreline Change

The shoreline position is measured from a reference line. The shoreline displacement is measured at discrete locations using a series of equidistant spaced (in this study 50 m), cross-shore transects. The methods are described by Dolan et al. (1978, 1980), Overton and Fisher (1996), and others, with similar methods used in Morton and Miller (2005).

Typically, shoreline location is measured along a cross-shore transects. Cross-shore transect spacing is decided by the scale of shoreline variability that is intended to be measured; Variability on a scale smaller than transect spacing contributes to uncertainty in the measurement. In this way, measuring shoreline along a transect is sensitive to transect placement. In this section, shoreline location is measured by calculating the area of land above MHW within the shoreline band. This is done for each 50 m wide segment. Then, the area is divided by the width of the segment giving an average displacement of the shoreline from the core shoreline. In averaging the shoreline measurement over the width of the segment, the uncertainty in the measurement due to small-scale variability is also divided by the segmented width. The use of area to measure shoreline location, as opposed to measuring along a single transect, attempts to reduce the uncertainty in the measurement resulting from transect placement. Beach area measurements are performed in a similar manner to volume measurements described in Sec. 5.3.

By applying line feature extraction to the core and envelope, the space within which the given line feature evolved can be mapped. For example, shorelines extracted from the core and envelope define a shoreline band within which the shoreline evolved during the given period (Fig. 4.9a). Additional metrics that provide quantitative information about mass redistribution within the evolving landscape have also been derived (Hardin et al. 2011).

Typically, coastal features run parallel to the shoreline; That is, they are measured along a series of cross-shore transects. We can find the locations where coastal

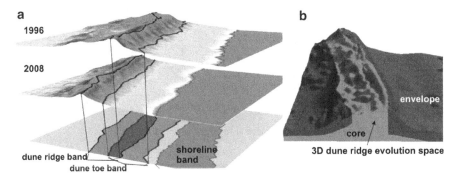

Fig. 4.9 Change in the dune ridge, dune toe and shoreline between the years 1998 and 2008: (**a**) respective evolution bands and (**b**) 3D ridge evolution space

features intersect transects using the v.clean command. Then we can compare
the changes in the location and characteristics of a feature across time steps, as
shown in the following GRASS code sample:

```
# Purpose: Extract baseline shoreline and generate
# cross-shore transects.
import grass.script as grass

grass.run_command( 'g.region', rast='NH_2008_1m')

grass.run_command( 'r.contour', input='NH_2008_1m',
    output='NH_2008_08m', levels=0.8, cut=600,
    overwrite=True )
grass.run_command( 'v.transects.py',
    input='NH_2008_08m', output='NH_transects',
    transect_spacing=50, dleft=50, dright=200,
    overwrite=True )
# Add columns to transects to hold intersection
# locations.
grass.run_command( 'v.category', input='NH_transects',
    output='temp', option='add', overwrite=True )
grass.run_command( 'g.rename',
    vect='temp,NH_transects', overwrite=True )
grass.run_command( 'v.db.addtable',
    map='NH_transects', table='table_trans', layer=1,
    columns='trans_num int,x_1999 double
    precision,y_1999 double precision,z_1999 double
    precision,x_2008 double precision,y_2008 double
    precision,z_2008 double precision,r2 double
    precision,del_z double precision' )
grass.run_command( 'v.db.update', map='NH_transects',
    col='trans_num', qcol='cat' )


# Use manually selected endpoints.
pt1="913861,250665"
pt2="914301,249736"
for date in [1999, 2008]:
    # Get dune ridge and intersection points.
    grass.run_command( 'r.mapcalc',
        expression='cost=exp(-5*NH_'+str(date)+'_1m)',
        overwrite=True )
    grass.run_command( 'r.cost', flags='k',
        input='cost', output='ccost', coordinate=pt1,
        stop_coordinate=pt2, null_cost=1, overwrite=True
        )
```

```
grass.run_command( 'r.drain', input='ccost',
    output='NH_'+str(date)+'_duneRidge',
    voutput='NH_'+str(date)+'_duneRidge',
    coordinate=pt2, overwrite=True )
grass.run_command( 'v.patch',
    input='NH_'+str(date)+'_duneRidge,NH_transects',
    output='temp', overwrite=True )
grass.run_command( 'v.clean', input='temp',
    output='temp_o',
    error='NH_'+str(date)+'_duneRidge_pts',
    tool='break', overwrite=True )
# Get transect number and elevation for each
# intersection point.
grass.run_command( 'v.category',
    input='NH_'+str(date)+'_duneRidge_pts',
    output='NH_'+str(date)+'_duneRidge_pts_cat',
    option='add', overwrite=True )
grass.run_command( 'v.db.addtable',
    map='NH_'+str(date)+'_duneRidge_pts_cat',
    table='table_'+str(date), layer=1,
    columns='trnsct int,z double precision' )
grass.run_command( 'v.what.vect',
    map='NH_'+str(date)+'_duneRidge_pts_cat',
    layer=1, column='trnsct', qmap='NH_transects',
    qlayer=1, qcolumn='trans_num', dmax=1 )
grass.run_command( 'v.what.rast',
    map='NH_'+str(date)+'_duneRidge_pts_cat',
    layer=1, raster='NH_'+str(date)+'_1m',
    column='z' )
# Populate transect table with dune data.
a = grass.read_command( 'v.out.ascii',
    input='NH_'+str(date)+'_duneRidge_pts_cat',
    output='-', columns='trnsct,z' )
a = a.strip('\n')
for line in a.split('\n'):
    line = line.split('|')
    grass.run_command( 'v.db.update',
        map='NH_transects', layer=1,
        column='x_'+str(date), value=line[0],
        where='trans_num='+line[2] )
    grass.run_command( 'v.db.update',
        map='NH_transects', layer=1,
        column='y_'+str(date), value=line[1],
        where='trans_num='+line[2] )
```

```
        grass.run_command( 'v.db.update',
           map='NH_transects', layer=1,
           column='z_'+str(date), value=line[4],
           where='trans_num='+line[2] )

grass.run_command( 'v.db.update', map='NH_transects',
   layer=1, column='r2',
   qcolumn='(x_2008-x_1999)*(x_2008-x_1999)
        +(y_2008-y_1999)*(y_2008-y_1999)' )
grass.run_command( 'v.db.update', map='NH_transects',
   layer=1, column='del_z', qcolumn='z_2008-z_1999' )
grass.run_command( 'v.db.select', map='NH_transects',
   columns='r2,del_z' )
```

4.6   Mapping Location and Change of Built Structures

We can map the location of homes in relatively flat coastal terrain using a simple raster analysis. For the case where $h_b = 9$ m, structures can be extracted by running the following GRASS commands:

```
# Purpose: Identify potential structures as elevations
# above threshold (9m) in envelope surface.
# Limit search to beach front by buffering shoreline.
g.region rast=NH_env
v.buffer input=NH_2008_08m \
   output=NH_2008_08m_200mbuff distance=200
v.to.rast input=NH_2008_08m_200mbuff \
   output=NH_2008_08m_200mbuff typ=area use=val value=1
r.mask raster=NH_2008_08m_200mbuff
# Extract the houses.
r.mapcalc \
   expression='NH_houses_rast=if((NH_env-9),1,null(),
   null())' --o
r.mask raster=NH_2008_08m_200mbuff -r
```

Beach-front homes extracted through raster analysis are shown in Fig. 4.10. Building footprint layers can be used to generate more complex cost surfaces for more accurate dune ridge extraction, e.g., by increasing the cost in building footprints:

```
# Purpose: Extract dune ridges.
r.mapcalc \
   expression='cost=exp(-2*if(isnull(NH_houses_rast), \
   NH_2008_1m, 9+(9-NH_2008_1m)))' --o
```

```
r.cost -k input=cost output=cumulative_cost \
   coordinate=913859,250658 \
   stop_coordinate=914305,249739 --o
r.drain input=cumulative_cost \
   output=NH_2008_duneRidge voutput=NH_2008_duneRidge \
   coordinate=914305,249739 --o
r.mapcalc\
 expression='NH_2008_duneRidge=float(\
 NH_2008_duneRidge)
 * NH_2008_1m' --o
```

Discrete terrain changes, such as construction or destruction of a building, are characterized by a significant difference in elevation for a set of grid cells measured between two time snapshots t_k and t_{k+1}. To accurately identify this type of change, adequate representation of structures is required. Usually this is found in multiple-return or last-return (not bare earth) lidar-derived DEMs with resolutions 0.3–0.5 m. Structures that were built or lost between the beginning and end of the study period

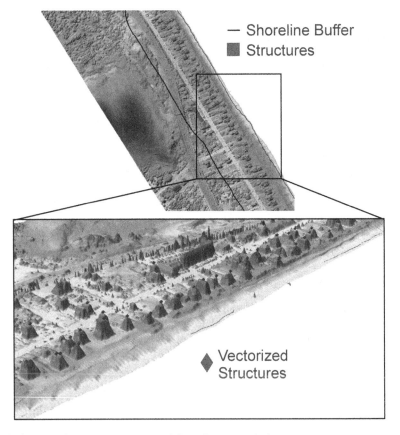

Fig. 4.10 Beach-front structures extracted through raster analysis

can be identified using the core and envelope surfaces, given by Eqs. (3.1) and (3.2) respectively, as grid cells (i_c, j_c) that fulfill the following condition:

$$z_{env}(i_c, j_c) - z_{core}(i_c, j_c) > h_b \qquad (4.13)$$

where h_b is the threshold relative height of the building captured by lidar. To map only the buildings that were built or lost, we can modify the map algebra expression above as follows:

```
# Purpose: Identify elevation difference between core
# and envelope above threshold (9 m).
g.region rast=NH_env
r.mask input=NH_2008_08m_200mbuff
r.mapcalc\
  expression='NH_houses_change=if((NH_env-NH_core)\
  >9,1,null(),null())'
r.mask input=NH_2008_08m_200mbuff -r
```

Lost structures will be located in grid cells (i_l, j_l) that fulfill the condition given by Eq. (4.13) and the following relation:

$$t_{max}(i_l, j_l) < t_{min}(i_l, j_l) \qquad (4.14)$$

while new structures can be identified as grid cells (i_n, j_n) where:

$$t_{max}(i_n, j_n) > t_{min}(i_n, j_n) \qquad (4.15)$$

with raster maps $t_{max}(i, j)$ and $t_{min}(i, j)$ defined by Eqs. (3.3) and (3.4). If more detailed temporal information is needed, the extracted new or lost buildings can be vectorized using the standard GIS tools and the associated centroids (i_c, j_c) can be used to perform an automated query of the entire DEM time series.

```
# Purpose: Identify structures that were either
# built or destroyed during the study period.
# Convert raster structures to vector areas.
r.to.vect input=NH_houses_rast output=NH_houses_vect \
    type=area --o
# Remove small areas, e.g, telephone poles.
v.clean input=NH_houses_vect \
    output=NH_houses_vect_clean tool=rmarea thresh=15 \
    --o
# Convert house centroid to point location.
v.extract input=NH_houses_vect_clean \
    output=NH_houses_centroid type=centroid --o
```

```
v.type input=NH_houses_centroid \
   output=NH_houses_point from_type=centroid \
   to_type=point --o
# Sample dem time series.
v.out.ascii input=NH_houses_point output=- \
   separator=' ' | r.what map=NH_1999_1m,NH_2001_1m
# Make a raster map showing built and lost homes
# in which
# positive/negative elevation differences indicate
# built/lost.
r.mapcalc expression='NH_elev_diff=(NH_env-NH_core) *\
   if(NH_2008_1m>NH_1999_1m,1,-1)'
v.db.addcolumn map=NH_houses_vect_clean columns="diff \
   DOUBLE PRECISION"
v.rast.stats map=NH_houses_vect_clean \
   raster=NH_elev_diff method=average col_prefix=mean
v.to.rast input=NH_houses_vect_clean \
   output=NH_houses_elev_diff use=attr \
   attrcolumn=mean_avera
# Use threshold of 3 to filter small differences.
r.mapcalc\
   expression="NH_houses_built_lost=if(\
   NH_houses_elev_diff>3,\
   1, if(temp2<-3,-1,null()))"
```

Elevation differences at grid cells (i_c, j_c) computed for the individual successive time snapshots t_p and t_{p+1}, $p = 1, \ldots, n-1$:

$$\Delta z(i_c, j_c, t_d) = z(i_c, j_c, t_p) - z(i_c, j_c, t_{p+1})$$

can be analyzed and a time interval t_d when a new house was built can be identified using condition:

$$\Delta z(i_c, j_c, t_d) < -h_b \qquad\qquad (4.16)$$

and when a house was lost can be identified with the following condition:

$$\Delta z(i_c, j_c, t_d) > h_b \qquad\qquad (4.17)$$

Using this approach, we can also investigate whether there were any homes that were built and quickly lost within the study period or that were lost and re-built (Fig. 4.11). Extracted information about the new and old buildings can be compared with county records to evaluate the results, verify permits for the new buildings and identify potential violations.

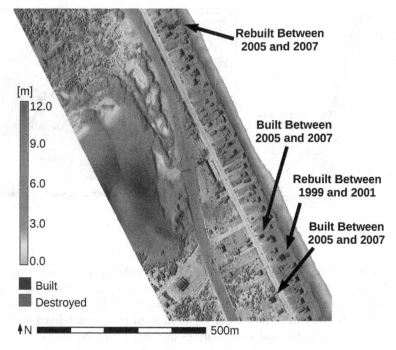

Fig. 4.11 Beach-front structures classified as built or destroyed

We can also extract a wide range of additional information about structures and their relation to terrain evolution to support decision making and coastal management. For example, we can identify vulnerable new structures that were built on a very small core (stand on moving sand) by combining Eqs. (4.13) and (4.15) with a condition:

$$z_{core}(i_n, j_n) < z_b \qquad (4.18)$$

where z_b is minimum core elevation considered safe (e.g. based on storm surge, or sea level rise). Homes located within the shoreline evolution band (already lost or highly vulnerable) can also be easily identified.

4.7 Derived Parameters: Storm Vulnerability Scale

Coastal vulnerability to storm events depends on storm characteristics as well as the continuously evolving, spatially varying beach and foredune topography (Morton 2002; Wright et al. 1970). Efforts to identify vulnerable locations have focused on characterizing the prestorm topographic features relative to storm parameters (e.g., García-Mora et al. 2001; Hallermeier and Rhodes 2011; Jiménez et al. 2009; Jin and

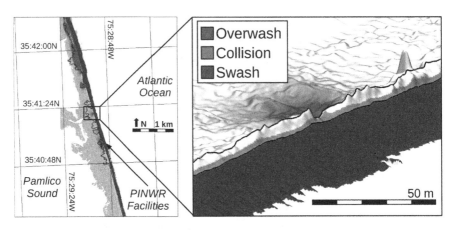

Fig. 4.12 Storm scale draped over DEM

Overton 2011; Judge et al. 2003; Sallenger 2000; Stockdon et al. 2007; Stockdon and Thompson 2007a,b). All of these efforts rely on topographic parameters (e.g., dune ridge height, dune toe height, beach width, and beach slope) that are typically extracted from cross-shore profiles or digitized from three-dimensional surfaces. The need to perform regional scale analyses over potentially hundreds of profiles has driven the development of automated techniques for extraction of critical coastal features and estimation of their parameters.

One of the vulnerability assessment methods that requires dune and beach parameters is the storm impact scale (Sallenger 2000). It specifies four distinct erosion regimes that can occur during storm events. These regimes are defined by the elevations of the pre-storm dune ridge D_{high} and the dune toe D_{low} relative to the storm surge elevation with and without run-up, R_{high} and R_{low}, respectively. Wave run-up is typically estimated using the 2 % exceedance levels and is a function of beach slope, $\tan \beta$, and the deep water wave height and period. The four distinct erosion regimes are as follows:

- The swash regime occurs when $R_{high} < D_{high}$ and is characterized by beach erosion and a likely post-storm recovery.
- The collision regime occurs when $D_{low} < R_{high} < D_{high}$ leading to dune erosion, with eroded sand unlikely to be redeposited back onto the dune after the storm.
- The overwash regime occurs when $R_{high} > D_{high}$ and is characterized by severe dune erosion and inland transport of sediment.
- Finally, the inundation regime occurs when $R_{low} > D_{high}$.

The inundation of a barrier island can flatten dunes, cause massive inland and offshore transport of sediment, and lead to barrier island breaching. A raster based implementation of the storm impact scale shown in Fig. 4.12 has been presented by Hardin et al. (2012).

References

Breen, D., House, D., and Getto, P. (1992). A physically-based particle model of woven cloth. the visual computer. *The Visual Computer*, 8(5-6):264–277. DOI: 10.1007/BF01897114.

Burroughs, S. and Tebbens, S. (2008). Dune retreat and shoreline change on the Outer Banks of North Carolina. *Journal of Coastal Research*, 24:104–112. DOI: 10.2112/05-0583.1.

Dolan, R., Hayden, B., and Heywood, J. (1978). A new photogrammetric method for determining shoreline erosion. *Coastal Engineering*, 2:21–39.

Dolan, R., Hayden, B., May, P., and May, S. (1980). The reliability of shoreline change measurements from aerial photographs. *Shore and Beach*, 48(4):22–29.

Douglas, D. and Peucker, T. (1973). Algorithms for the reduction of the number of points required to represent a digitized line or its caricature. *The Canadian Cartographer*, 10(2):112–122.

Ebisch, K. (2002). A correction to the Douglas–Peucker line generalization algorithm. *Computers & Geosciences*, 28:995–997.

Elko, N., Sallenger, A., Guy, K., Stockdon, H., Karen, L., and Morgan, L. (2002). Barrier island elevations relevant to potential storm impacts: 1. techniques. Technical report, US Geological Survey.

García-Mora, M., Gallego-Fernández, J., Williams, A., and Garcia-Novo, F. (2001). A coastal dune vulnerability classification. A case study of the SW Iberian Peninsula. *Journal of coastal research*, pages 802–811.

Hallermeier, R. and Rhodes, P. (2011). Generic treatment op dune erosion for 100-year event. *Proceedings of the International Conference on Coastal Engineering*, 1(21).

Hardin, E., Kurum, M., Mitasova, H., and Overton, M. (2012). Least cost path extraction of topographic features for storm impact scale mapping. *Journal of Coastal Research*.

Hardin, E., Mitasova, H., and Overton, M. (2011). Quantification and Characterization of Terrain Evolution in the Outer Banks, N.C. In *Proceedings of the Coastal Sediments '11, Miami, FL*, pages 739–753.

Hershberger, J. and Snoeyink, J. (1992). Speeding Up the Douglas-Peucker Line-Simplification Algorithmn. In *Proceedings of the 5th Symposium on Data Handling*, pages 134–143.

Jiménez, J., Ciavola, P., Balouin, Y., Armaroli, C., Bosom, E., and Gervais, M. (2009). Geomorphic coastal vulnerability to storms in microtidal fetch-limited environments: Application to NW Mediterranean & N Adriatic Seas. *J. Coast. Res., SI*, 56:1641–1645.

Jin, Q. and Overton, M. (2011). Quantitative analysis of coastal dune erosion based on geomorphology features and model simulation. In *Proceedings of the Coastal Sediments '11, Miami, FL*, pages 1825–1840.

Judge, E., Overton, M., and Fisher, J. (2003). Vulnerability indicators for coastal dunes. *Journal of Waterway, Port, Coastal, and Ocean Engineering*.

Mitasova, H., Drake, T., Bernstein, D., and Harmon, R. (2004). Quantifying rapid changes in coastal topography using modern mapping techniques and geographic information system. *Environmental and Engineering Geoscience*, 10:1–11. DOI: 10.2113/10.1.1.

Mitasova, H., Hardin, E., Starek, M., Harmon, R., and Overton, M. (2011). Landscape dynamics from LiDAR data time series. *Geomorphometry 2011, Redlands, CA*, pages 3–6.

Mitasova, H. and Hofierka, J. (1993). Interpolation by regularized spline with tension: II. Application to terrain modeling and surface geometry analysis.

Mitasova, H., Hofierka, J., Zlocha, M., and Iverson, L. (1996). Modelling topographic potential for erosion and deposition using GIS. *International Journal of Geographical Information Systems*, 10(5):629–641.

Mitasova, H., Overton, M., Oliver, R., and Hardin, E. (2012). Ocean shoreline migration. Technical report, Albemarle-Pamlico National Estuary Program.

Morton, R. (2002). Factors controlling storm impacts on coastal barriers and beaches – a preliminary basis for near real-time forecasting. *Journal of Coastal Research*, 18:486–501.

Morton, R. and Miller, T. (2005). NC_TRANSECTS_ST - short-term shoreline change rates for north carolina atlantic coast generated at a 50m transect spacing, 1970-1997. Technical report, U.S. Geological Survey Open-File Report 2005-1326.

Overton, M. and Fisher, J. (1996). Shoreline analysis using digital photogrammetry. In *Coastal Engineering (1996)*, pages 3750–3761. ASCE.

Provot, X. (1995). Deformation constraints in a mass-spring model to describe rigid cloth behaviour. In *Graphics Interface 95*, pages 147–154, Quebec, Canada.

Sallenger, A. (2000). Storm impact scale for barrier island. *Journal of Coastal Research*, 16:890–895. ISSN: 0749-0208.

Stockdon, H., Sallenger, A., and Holman, R. (2007). A simple model for the spatially-variable coastal response to hurricanes. *Marine Geology*, 238:1–20. DOI: 10.1016/j.margeo.2006.11.004.

Stockdon, H. and Thompson, D. (2007a). Vulnerability of National Park Service beached to inundation during a direct hurricane landfall: Cape Lookout National Seashore. *U.S. Geological Survey Open-File Report 2007-1376*.

Stockdon, H. and Thompson, D. (2007b). Vulnerability of National Park Service beached to inundation during a direct hurricane landfall: Fire Island National Park. *U.S. Geological Survey Open File Report 2007-1389*.

Tarboton, D. G. (1997). A new method for the determination of flow directions and upslope areas in grid digital elevation models. *Water resources research*, 33(2):309–319.

Weaver, K., Mitasova, H., and Overton, M. (2010). Geospatial analysis of the dynamics of a coastal sand dune using time series of lidar data and tangible geospatial modeling system (tangeoms). *Evolution*, 20:22.

Wright, L., Swaye, F., and Coleman, J. (1970). Effects of Hurricane Camille on the landscape of the Breton-Chandeleur Island chain and the eastern portion of the lower Mississippi delta. Technical report, DTIC Document.

Chapter 5
Volume Analysis

Coastal landform change is driven by sediment transport and redistribution of sand. In this chapter, we present techniques for mapping volumes of land mass using rectangular segments and analyzing volume evolution and redistribution in absolute and relative terms.

5.1 DEM Differencing

DEMs can be differenced to produce a map that represents the change in the elevation surface between the two time snapshots. These maps are sometimes referred to as DEMs of Difference (DoD) and can be produced in GRASS using r.mapcalc. For example, the total change in elevation within the Nag's Head study area from the beginning of the study period (1999) to the end of the study period (2008), which is shown in Fig. 5.1, can be can be computed by running the following GRASS commands:

```
# Purpose: Compute a DEM of Difference (DoD).
r.mapcalc \
    expression='NH_total_change=NH_2008_1m-NH_1999_1m'
r.colors map=NH_total_change \
    rules=color_elevation_diff.txt
```

Volume change per raster cell is then obtained by multiplying the elevation change by the raster cell area.

© The Author(s) 2014
E. Hardin et al., *GIS-based Analysis of Coastal Lidar Time-Series*, SpringerBriefs in Computer Science, DOI 10.1007/978-1-4939-1835-5_5

Fig. 5.1 DEMs representing the terrain at the beginning (1999) and end (2008) of the study period, as well as the difference between the two maps, representing the terrain change observed during the student period

5.2   Landscape Segmentation into Bins

To compute volumes and volume change for areas larger than the raster cell (to reduce noise and provide information more indicative of the local coastal state) but still small enough to provide information about spatial redistribution of sand, the beach-foredune area can be partitioned into the rectangular segments. These segments can be generated by combining long-shore partitions with cross-shore transects.

5.2.1   Long-Shore Partitioning

We have already derived the *core* and *envelope* surfaces as the minimum and maximum elevations measured for each raster cell. We have also delineated a shoreface area, called the *shoreline band*, as the area between the MHW contours of the core and envelope surfaces. The area within the shoreline band bounds shoreline evolution over the study period, and the width of the shoreline band measures the shoreline migration range (Mitasova et al. 2012). The shoreline band will be our first long-shore partition.

The second long-shore partition is defined inland of the shoreline band. It is bound by the core (minimum) shoreline and by a horizontal distance of 110 m inland of the core shoreline to bound the upper-beach dune section. The constant inland distance of 110 m was chosen to ensure complete lidar data coverage for each year. The area within the shoreline band and the area that extends inland of the shoreline band (Fig. 5.2) can be extracted using r.mapcalc in conjunction with v.buffer:

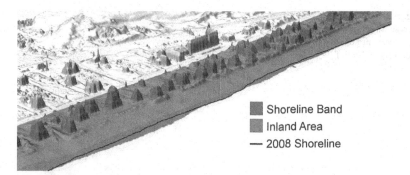

Fig. 5.2 Area within the shoreline band (extracted as the area bounded by the core and envelope shorelines) and the area that extends 110 m inland of the shoreline

```
# Purpose: Extract area within shoreline band.
r.mapcalc \
    expression='shorelineBand=if(isnull(NH_core) && \
    !isnull(NH_env), 1, null())'
# Extract area landward of the shoreline band.
r.contour input=NH_core output=NH_core_shoreline \
    level=0.8 cut=400
v.buffer input=NH_core_shoreline \
    output=NH_core_shoreline_110mbuff distance=110
v.to.rast input=NH_core_shoreline_110mbuff \
    output=NH_core_shoreline_110mbuff typ=area use=val \
    value=1
r.mapcalc expression='coreArea=if(!isnull(NH_core) && \
    !isnull(NH_core_shoreline_110mbuff), 1, null())'
```

The area inland of the core shoreline will be referred to as the *area with core* as opposed to the *shoreline band area* which has no core surface in it.

5.2.2 Cross-Shore Segments

In order to quantify along-shore trends, we can further partition the beach foredune system using cross-shore transects perpendicular to a baseline. The baseline can be a selected shoreline (in our example we use the core surface shoreline) or an off-shore line approximately parallel with a selected shoreline. Cross-shore segments can be generated using the GRASS add-on module v.transects, by setting the type parameter to area.

Fig. 5.3 Vector map
representing the volumes for
each segment area displayed
over a shaded relief

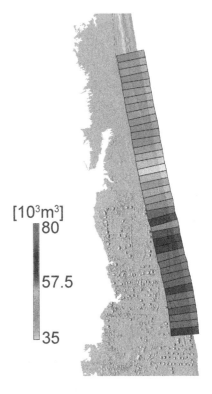

```
# Purpose: Segment the DEM into cross-shore areas
# along the shoreline.
r.contour input=NH_core output=NH_core_shoreline\
   level=0.8 cut=400 --o
# v.clean and v.build.polylines were required to keep
# the contour from doubling back
v.clean input=NH_core_shoreline \
   output=NH_core_shoreline_c tool=rmsa --o
v.build.polylines input=NH_core_shoreline_c \
   output=NH_core_shoreline --o
v.transects.py input=NH_core_shoreline \
   output=NH_alongShoreSegments dleft=50 dright=150 \
   type=area transect_spacing=50 --o
```

An overlay of long-shore and cross-shore segments will create partitioning of the beach-foredune system into rectangular segments as shown in Fig. 5.3. These partitions will allow us to map and quantify the volume change along the highly dynamic shoreline band and along the upper beach-foredune segments.

5.3   Volume Estimation for Segments

Volume of mass over a given area can be estimated from a DEM by summing elevations in raster cells defining this area

$$V = A \sum_{i} \sum_{j} z(i, j) \tag{5.1}$$

where V is the volume, A is the area of a raster cell, z is the elevation of a surface, and i and j are summed over all raster cells in the area for which volume is computed. In the following example, we estimate the total volume defined by a raster surface and horizontal plane or datum over a given area using the `r.volume` command. We set a MASK to the raster map `coreArea` to limit the volume calculation to the area of interest.

```
# Purpose: Limit volume measurement to 100 m of
# shoreline.
r.mask input=coreArea
r.volume NH_2008_1m
r.mask input=coreArea -r
```

   To calculate the volume for each cross-shore segment (which is a vector area), we connect a table to the vector map. Then we populate the table with statistics derived from the DEM using the `v.rast.stats` command as in the following GRASS code:

```
# Purpose: Calculate the volume in each segment.
v.db.addtable map=NH_alongShoreSegments
v.rast.stats vector=alongShoreSegments \
    raster=NH_2008_1m column_prefix=NH2008
v.db.addcolumn map=NH_alongShoreSegments \
    column='volume DOUBLE PRECISION'
# volume = raster cell area (which is 1) * sum of
# elevations
v.db.update map=NH_alongShoreSegments col=volume \
    qcol="1*NH2008_sum"
v.db.select map=NH_alongShoreSegments \
    columns="cat,volume"
```

We can then display the volumes as colored vector areas using the module `d.vect.thematic` (Fig. 5.3).

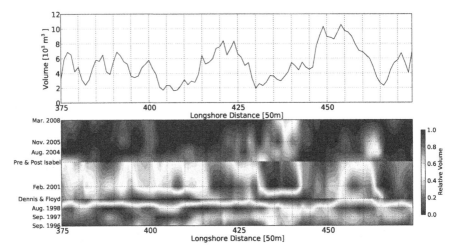

Fig. 5.4 Graph showing volumes in the shoreline band and a heat map showing volume evolution

5.4   Volume Change Metrics

Given the volume maps computed for each time snapshot, we can analyze the pattern of volume change. In addition to a standard graph showing volumes along the segments for each year (Fig. 5.4), we can compute the volume differences between any given time snapshots for all segments using vector attributes database operations or by converting the vector volumes map to raster representation and using map algebra. Evolution of volumes over time can be represented using a heat map (Tateosian et al. 2013) (Fig. 5.4).

Relative volumes, normalized according to the volume of the dynamic layer, allow us to analyze the volume of each segment relative to the minimum (core) and maximum (envelope) volume and how this pattern has changed over time. Relative volume in the upper beach—foredune area (area with core, inland from the core shoreline) is defined for each time snapshot and for each segment j as follows:

$$\hat{V}_{ij} = \frac{V_{ij} - V_{cj}}{V_{ej} - V_{cj}},\tag{5.2}$$

where $\hat{V}_{ij}$ is the relative volume for the ith survey in the time series, V_{ij} is the volume under the ith elevation surface, V_{ej} is the volume under the envelope surface, and V_{cj} is the volume under the core surface. Volumes were calculated relative to MHW. This relative volume can then be calculated by running the following GRASS commands:

```
# Purpose: Calculate relative volume.
r.mask input=coreArea
r.mapcalc\
  expression="NH_2008_volRel_inland=\
  (NH_2008_1m-NH_core)/(NH_env-NH_core)"
r.univar NH_2008_volRel_inland
 [..]
 mean: 0.442037
 [..]
r.mask input=coreArea -r
```

Relative volume within the shoreline band (between the core and envelope shorelines) is defined for each time snapshot and for each segment j as follows:

$$\hat{W}_{ij} = \frac{W_{ij}}{W_{ej}}, \tag{5.3}$$

where $\hat{W}_{ij}$ is the relative volume for the ith survey in the time series, W_{ij}, is the volume under the ith elevation surface, and W_{ej} is the volume under the envelope surface within the shoreline band. By definition, the core surface does not exist within the shoreline band above MHW and therefore the core volume W_{cj} is equal to zero. Relative volume in the shoreline band can be calculated by running the following commands:

```
# Purpose: Calculate relative volume in the shoreline
# band.
r.mask input=shorelineBandr
r. mapcalcexpression=\
 "NH_2008_volRel_shoreband=NH_2008_1m/NH_env"
r.univar NH_2008_volRel_shoreband
 [..]
 mean: 0.863858
 [..]
r.mask input=shorelineBand -r
```

Although it is typical to report volumetric analysis in absolute values in units of m³ (Burroughs and Tebbens 2008; White and Wang 2003), analyzing and visualizing relative volume offers some advantages: First, the core surface gives a lower bound on terrain evolution, and for this reason, the core is a logical datum. Removal of the core values from the analysis highlights changes, (e.g., in areas where the volume of transported sediment is much less than the volume of the stable sediment under the core surface). Second, because the core represents a minimum bound on volume evolution, volumes near the core volume represent worst case scenarios observed in the time-series. Finally, because the terrain evolved exclusively within the dynamic layer, visualizing volume as a percent of the dynamic layer volume allows for an

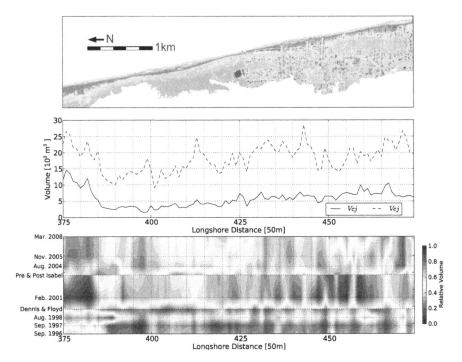

Fig. 5.5 Graph showing core and envelope volumes, and heat map showing relative volume evolution

at-a-glance determination of the present state relative to the minimum and maximum observed over the study period. Evolution of the relative volumes can be visualized using a heat map (Fig. 5.5).

References

Burroughs, S. and Tebbens, S. (2008). Dune retreat and shoreline change on the Outer Banks of North Carolina. *Journal of Coastal Research*, 24:104–112. DOI: 10.2112/05-0583.1.

Mitasova, H., Overton, M., Oliver, R., and Hardin, E. (2012). Ocean shoreline migration. Technical report, Albemarle-Pamlico National Estuary Program.

Tateosian, L., Mitasova, M., Thakur, S., Hardin, E., Russ, E., and Bruce, B. (2013). Visualizations of coastal terrain time-series. *Information Visualization*, 13:266–282.

White, S. and Wang, Y. (2003). Utilizing DEMs derived from LIDAR data to analyze morphologic change in the North Carolina coastline. *Remote Sensing of Environment*, 85(1):39–47. DOI: 10.1016/S0034-4257(02)00185-2.

Chapter 6
Visualizing Coastal Change

Scientific visualization provides a means for effective analysis and communication of complex information that may be otherwise difficult to explain and explore. This particularly applies to coastal geomorphology, where 3D spatial and temporal patterns and relationships are critical for capturing landscape features and their dynamics. In this chapter we present GIS-based techniques for visualizing dynamic coastal landscapes using 2D maps, 3D perspective views, animations, and the space-time cube approach.

6.1 Color and Relief Shading

Continuous-tone color ramps are commonly used to represent elevation, elevation change, and topographic parameters. The colors can be assigned with equal intervals or with variable intervals based on the statistical distribution of the mapped values. For example, histogram equalized color ramps employ a monotonic, non-linear mapping which assigns the color values to grid cells to achieve a uniform color distribution. These color ramps are effective for highlighting topographic features in regions with an uneven distribution of mapped values (Fig. 6.1). Divergent color ramps are used for variables which have positive and negative values indicating an opposing property, such as, gain and loss (Fig. 5.1) or convexity and concavity (negative and positive curvatures, Fig. 6.2). Discrete color maps are best suited for the classified features, such as new and old homes, discrete maps, such as time of minimum (Fig. 3.2) but they are also effective for continuous features when highlighting certain intervals or if a simplified representation of the spatial pattern is needed. The color tables for GRASS GIS raster maps are managed using the `r.colors` command. You can find most of the custom colors tables used in this book in the Appendix, formatted as a text input for the `r.colors` command.

© The Author(s) 2014
E. Hardin et al., *GIS-based Analysis of Coastal Lidar Time-Series*, SpringerBriefs
in Computer Science, DOI 10.1007/978-1-4939-1835-5_6

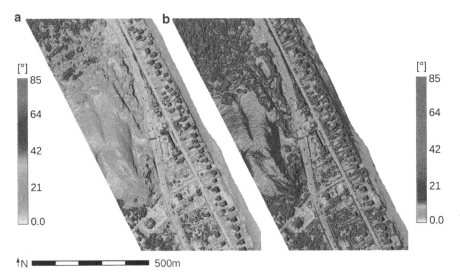

Fig. 6.1 (a) Equal interval and (b) histogram equalized color ramps for a slope map

Fig. 6.2 Curvature map overlaid on Jockey's Ridge DEM

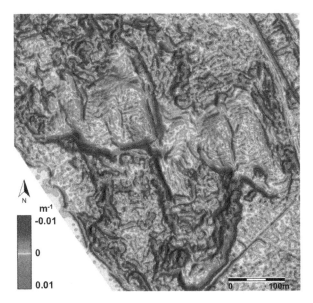

Relief shading (Horn 1981) combined with color is gradually replacing contours as a means for representing and analyzing elevation data. This technique captures subtle terrain features that are often missed by the traditional contour representation. It has become one of the preferred techniques for visualization of high resolution lidar-based DEMs. Images representing relief shading are derived by computing the image intensity values as a function of the illumination angle θ (the angle between the incoming light source ray and elevation surface normal):

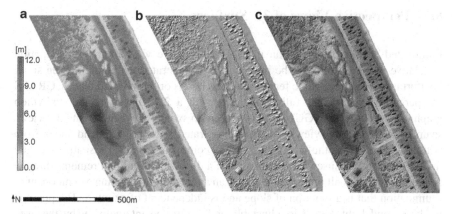

Fig. 6.3 Creating a colored relief shaded map: (**a**) elevation color image (**b**) shaded relief image (**c**) combined elevation color as hue and shaded relief as intensity

$$\cos(\theta) = \cos(\omega)\cos(\gamma) + \sin(\omega)\sin(\gamma)\cos(\beta - \alpha) \qquad (6.1)$$

where ω is light source altitude angle measured from the zenith, γ is the elevation surface slope angle, β is the light source azimuth, and α is the elevation surface aspect. Landscapes with the relatively flat topography common in coastal regions, require a large ω angle (low light source) to reveal shallow depressions, narrow foredunes, and other types of small features. We compute the image representing illuminated topography with the r.shaded.relief command. In coastal regions, exaggerating relief by setting the zmult parameter to 3 further improves the visualization of subtle landforms. The resulting map is then displayed by the d.his command with relief shading used as intensity and elevation color used as hue:

```
# Purpose: Compute shaded relief map.
r.shaded.relief input=NH_2008_1m output=NH_2008_shade \
    zmult=3
d.his i=NH_2008_shade h=NH_2008_1m
```

The DEM and the map representing relief shading are shown in Fig. 6.3.

Relief shading provides a 2D orthogonal view of the topography at a uniform scale and it is suitable for landform mapping using on-screen digitizing. Several add-ons modules, such as r.shaded.pca, r.local.relief, r.skyview, and r.sun.daily provide additional capabilities for visualization of topography based on surface illumination from multiple directions or terrain openness (see the manual pages for these modules for more details and references).

6.2   Perspective Views of 3D Surfaces

Illuminated surface visualization in a 3D perspective view improves perception
of relative elevation that can be interactively exaggerated to highlight even small
landforms and anthropogenic features such as berms or buried sand fences. GRASS
3D perspective viewer is fully integrated with a 2D display in the wxPython
graphical user interface (GUI). It allows users to switch between 2D and 3D views
seamlessly. Perspective views can also be generated from the command line or from
within a script using the m.nviz.image command. The 3D visualization tool
in GRASS wxnviz uses two light sources: a dim white light that remains directly
above the surface at all times and serves as an ambient light, creating a component of
illumination that is a function of slope and is independent of azimuth. The position
of the second light source is adjustable and controlled interactively by the user.
When light is being adjusted, a sphere appears on the surface and is continuously
redrawn to show the effects of lighting changes (Fig. 6.4).

A color map draped over an elevation model is widely used to convey the
relationship between the surface geometry and parameters derived from the DEM,
such as slope, aspect, or curvatures (Fig. 6.2). Line, point, or polygon features can be
draped over the illuminated elevation surface to provide baseline information, such
as roads and building footprints, or geomorphologic features such as shorelines,
ridge lines, or peaks (Figs. 4.4 and 4.6). Surfaces can also be combined with 3D
vector data. For example, to explore the relationship of a multiple return point
cloud with the bare ground surface or to represent structures such as buildings
and bridges. Several screen capture videos demonstrate the 3D visualization tool,
for example, http://courses.ncsu.edu/mea582/common/GIS_anal_grass/GIS_Anal_
grvisual.html.

6.3   Comparing Multiple Surfaces: Map Swipe
and 3D Cross-Sections

Changes in elevation and land cover between two time snapshots can be visually
compared using a slider tool, also referred to as mapswipe (Fig. 6.5), available through wxGUI (http://grasswiki.osgeo.org/wiki/WxGUI_Map_Swipe) or as
a command g.gui.mapswipe.

Overlaid multiple surfaces visualized in 3D perspective view with interactive
cross-sections are effective for analyzing topographic change and land surface
evolution. A reference plane at constant elevation improves the perception of relative positions between surfaces in cross-sections. Screen capture video "Visualization in GRASS GIS III: cutting planes" (http://courses.ncsu.edu/mea582/common/
GIS_anal_grass/GIS_Anal_grvisual.html) demonstrates the interaction with cutting
planes to create the cross-section application for our coastal data set is shown by
Fig. 3.1.

Southeast directed source

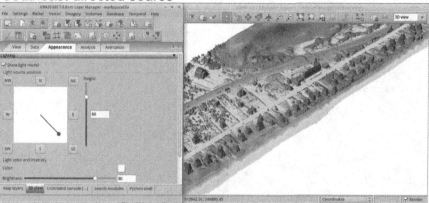

Southwest directed source

Fig. 6.4 Interactively changing the light source direction in the wxPython GUI. The lighting on the appearance of the DEM is apparent

6.4 Animations in 2D and 3D Space

Animations have become an indispensable tool for analyzing and visualizing time series of coastal terrain monitoring data and models of terrain evolution. Sequences of raster maps computed by simulation tools or interpolated lidar survey data can be animated using a series of 2D images or perspective views (Mitas et al. 1997). A series of color maps sequentially draped over a 3D perspective view of a static elevation surface and evolving 3D surfaces are effective approaches for analyzing and communicating the relationship between landforms and process dynamics. Simultaneous animation of 2D images and 3D perspective views, together with vector data is supported by the animation tool available through wxGUI (http://grasswiki.osgeo.org/wiki/WxGUI_Animation_Tool) or as a command g.gui.animation.

Fig. 6.5 Using the slider to compare two images, 1998 and 2008 DEMs

6.5   Visualization with Space-Time Cube (STC)

The space-time cube (STC) approach plots spatio-temporal data within a reference cube, where the xy-plane represents geographic position and the vertical axis (z-direction) represents time. STC is effective for visualizing trajectories of objects and movement data (Kristensson et al. 2009; Kwan and Lee 2004; Shrinivasan 2005), multivariate time-varying data (Li and Kraak 2005; Tateosian et al. 2013), and discrete data derived from DEMs (Thakur et al. 2013).

We can stack a series of DEMs in an STC to create a voxel model of terrain evolution. We can then extract an isosurface for a selected elevation value to represent terrain evolution along this elevation contour. For example, the shoreline contours, when displayed as a 2D map result in a tangle of lines that may be difficult to interpret (Tateosian et al. 2013). As an alternative to these overlapping lines, the shoreline can be extracted from the voxel model to create an isosurface representing shoreline evolution (Starek et al. 2011). To enhance interpretation of such isosurfaces we drape a color map over the isosurface to associate the stratum of the isosurface with the epoch (time period). Alternatively, other attributes, such as the rate of change or distance to a road, can also be represented by a draped

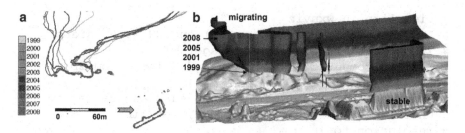

Fig. 6.6 Evolution of a small migrating dune along elevation contour: (**a**) displayed as a set of overlapping contours in 2D and (**b**) represented as an isosurface extracted from a space-time cube voxel model

color map (Starek et al. 2011). In the following example, we stack a series of lidar-based DEMs into a voxel model and visualize the evolution of elevation contours associated with a small migrating dune as isosurfaces (Fig. 6.6):

```
# Purpose: Visualize terrain dynamics as space-time
# cube.
# Set the 3D region.
g.region n=250416 s=249942 w=913734 e=914022 t=40 \
   b=12 \
       nsres=2 ewres=2 res3=4 tbres=4 -ap3
# Stack the DEMs into voxel model
# (ignoring variable time interval in this example).
r.to.rast3 \
   in=NH_1999_1m,NH_2001_1m,NH_2004_1m,NH_2005_1m,
      \NH_2007_1m,NH_2008_1m \
      out=NH_99_08_stack
r3.info NH_99_08_stack
r3.univar NH_99_08_stack
r3.stats NH_99_08_stack nsteps=10

# Create a volume for isosurface coloring according
# to time.
r.mapcalc NH_1999_t=1999
r.mapcalc NH_2001_t=2001
r.mapcalc NH_2004_t=2004
r.mapcalc NH_2005_t=2005
r.mapcalc NH_2007_t=2007
r.mapcalc NH_2008_t=2008
r.to.rast3 \
   input=NH_1999_t,NH_2001_t,NH_2004_t,NH_2005_t,
      \NH_2007_t,NH_2008_t \
      out=NH_9908_t
```

```
# Visualize in wxnviz
# Add NH_2008_1m DEM (2D raster) and NH_99_08_stack
# (3D raster) in GIS layer manager
# Zoom to computational region and switch from 2D to
# 3D view
# Under the View tab: set view to east and z-exag to 3
# Under the Data tab: for Surface, set fine
# resolution to 1
# Under the Data > Volume tab select the
# NH_99_08_stack as 3D raster map
# (it should be there already)
# Select Draw Mode > isosurfaces, Resolution 1
# For List of isosurfaces Add > Level 9
# by typing 9 for Isosurface attributes > Isosurface
# value
# for isosurface color map you can keep the default
# or select NH_9908_t to color the isosurface
# according to the years
# Adjust the vertical position of the volume to
# around 10
# to pull it above the DEM
# Adjust the view under View tab and lighting under
# Appearance tab
# You can add isosurfaces with additional values
# or switch to crossections by selecting Mode slices
# after saving your settings in wxnviz
# you can use m.nviz.image to generate the images
# through command line
```

To apply this approach to a time series of point clouds, we define a trivariate function, G, to represent land surface evolution:

$$z = G(x, y, t) \tag{6.2}$$

where x, y is the horizontal location, t is the time coordinate and elevation z is the modeled variable. The function $G(x, y, t)$ can be derived from a series of m point clouds $\{(x_i, y_i, z_i), i = 1, \ldots n_k\} t_k, k = 1, \ldots, m$ where (x, y, z) are coordinates, n_k is number of points in the kth point cloud and t_k is the time of the survey. We merge the data from all point clouds and re-organize them into a single point cloud $(x_i, y_i, t_i, z_i) i = 1, \ldots, \sum n_k$ that is then interpolated into a voxel model (3D grid) using a trivariate interpolation function, the regularized smoothing spline, with anisotropic tension applied in the time dimension (Mitasova et al. 1995). Oct-tree segmentation is used to support spatial interpolation of the large merged point cloud. Time resolution is selected to be close to the time interval of the surveys, although the approach is designed to handle irregular time intervals as well. Evolution of a given contour $z = c$ is then visualized as a set of isosurfaces extracted from the voxel model. For example, shoreline evolution will be represented by the isosurface $z = z_{MHW}$, where z_{MHW} is the mean high water elevation level.

```
# Purpose: Interpolate the volume from the
# given point data and
# (optionally) compute gradients.
# Time is given in 100-day units (400-day resolution)
# give the space-time volume box approximately same
# values as our DEM.

v.in.ascii -z JR9908_xytz_100d.txt \
   out=JR9908_xytz_100d x=1 y=2 z=3
v.info JR9908_xytz_100d
g.region n=250416 s=249942 w=913734 e=914022 t=40 \
   b=12 nsres=2 ewres=2 res3=4 tbres=4 -ap3
v.vol.rst in=JR9908_xytz_100d wcol=dbl_4 ten=40 \
   smo=0.5 segm=30 npmin=250 dmin=5 \
   elev=JRxytz9908_4m100d

# you can visualize the JRxytz9908_4m100d voxel model
# using wxnviz
# as in our previous example and you can check the
# interpolated time slices as follows
r3.to.rast JRxytz9908_4m100d out=JRxytz9908_4m100d

# browse the resulting series of maps using wxgui
# animation tool
```

Visual analysis of space-time isosurface topology is useful for identifying specific surface evolution features. For example, if the extracted contour represents elevation close to a foredune ridge, "holes" in the isosurface represent temporal loss of elevation that has recovered, typical for an overwash after which the dune was repaired or recovered (Starek et al. 2011). For additional examples of STC application to elevation data see Mitasova et al. 2012; Starek et al. 2011, and Tateosian et al. 2013.

References

Horn, B. (1981). Hill shading and the reflectance map. *Proceedings of the IEEE*, 69(1):14–47.
Kristensson, P. O., Dahlback, N., Anundi, D., Bjornstad, M., Gillberg, H., Haraldsson, J., Martensson, I., Nordvall, M., and Stahl, J. (2009). An evaluation of space time cube representation of spatiotemporal patterns. *Visualization and Computer Graphics, IEEE Transactions on*, 15(4):696–702.

Kwan, M.-P. and Lee, J. (2004). Geovisualization of human activity patterns using 3D GIS: a time-geographic approach. *Spatially integrated social science*, 27.

Li, X. and Kraak, M.-J. (2005). New views on multivariable spatiotemporal data: the space time cube expanded. In *International Symposium on Spatio-temporal Modelling, Spatial Reasoning, Analysis, Data Mining and Data Fusion*, volume 36, pages 199–201.

Mitas, L., Brown, W., and Mitasova, H. (1997). Role of dynamic cartography in simulations of landscape processes based on multivariate fields. *Computers & Geosciences*, 23(4):437–446.

Mitasova, H., Harmon, R. S., Weaver, K. J., Lyons, N. J., and Overton, M. F. (2012). Scientific visualization of landscapes and landforms. *Geomorphology*, 137(1):122–137.

Mitasova, H., Mitas, L., Brown, W. M., Gerdes, D. P., Kosinovsky, I., and Baker, T. (1995). Modelling spatially and temporally distributed phenomena: new methods and tools for GRASS GIS. *International Journal of Geographical Information Systems*, 9(4):433–446.

Shrinivasan, Y. (2005). Visualization of spatio-temporal patterns in public transport data.

Starek, M. J., Mitasova, H., Hardin, E., Weaver, K., Overton, M., and Harmon, R. S. (2011). Modeling and analysis of landscape evolution using airborne, terrestrial, and laboratory laser scanning. *Geosphere*, 7(6):1340–1356.

Tateosian, L., Mitasova, M., Thakur, S., Hardin, E., Russ, E., and Bruce, B. (2013). Visualizations of coastal terrain time-series. *Information Visualization*, 13:266–282.

Thakur, S., Tateosian, L., Mitasova, H., Hardin, E., and Overton, M. (2013). Summary visualizations for coastal spatial-temporal dynamics. *International Journal for Uncertainty Quantification*, 3(3).

Appendix

1 Sample Datasets

This appendix provides a description of the data used in this book and attribution to the original source.

Nags Head Lidar
Naming Convention: `NH_*_lidar.txt`
Source: Digital Coast - NOAA Coastal Services Center Website: http://www.csc.noaa.gov/digitalcoast/
Data Type: Lidar
Purpose: To provide elevation data for a portion of the town of Nags Head.

Rodanthe Lidar
Naming Convention: `R_*_lidar.txt`
Source: Digital Coast - NOAA Coastal Services Center Website: http://www.csc.noaa.gov/digitalcoast/
Data Type: Lidar
Purpose: To provide elevation data for a portion of the town of Rodanthe and some of the Pea Island Wildlife Refuge.

Highway NC 12 Centerline
Naming Convention: `road_centerline.txt`
Source: Original to this book
Website: http://geospatial.ncsu.edu/osgeorel/data.html Data Type: Manually digitized points
Purpose: To provide points along the centerline of highway NC 12 within the town of Nags Head to be used for systematic error correction.

© The Author(s) 2014
E. Hardin et al., *GIS-based Analysis of Coastal Lidar Time-Series*, SpringerBriefs in Computer Science, DOI 10.1007/978-1-4939-1835-5

Highway NC 12 Lidar

Naming Convention: `DARE_BE94zm3_01m_rstdm.txt`
Source: North Carolina Department of Public Safety
Website: https://www.ncdps.gov/
Data Type: Lidar
Purpose: To provide elevation of highway NC 12 within the town of Nags Head to
be used for systematic error correction.

2   Color Tables

This appendix provides color tables used throughout the book.
`color_elev_coast.txt`

```
-2 aqua
-.2 aqua
-0.1 grey
0.1 grey
1 yellow
3 orange
5 green
10 brown
40 white
```

`color_stddev.txt`

```
0    245 245 220 #beige
0.1 193 255 193 #darkseagreen1
0.25 180 238 180 #darkseagreen2
0.5 155 205 155 #darkseagreen3
0.75 105 139 105 #darkseagreen4
1    255 165 0    #orange1
1.25 238 154 0    #orange2
1.5 205 133 0    #orange3
1.75 139 90 0    #orange4
2    255 127 0 #darkorange 1
2.25 238 118 0 #darkorange 2
2.5 205 102 0 #darkorange 3
2.75 139 69 0 #darkorange 4
3 255 69 0 #orangered 1
3.25 238 64 0    #orangered 2
3.5 205 51 51    #brown 3
4 139 35 35 #brown 4
7.3 220 20 60    #crimson
```

color_range.txt

```
 0 255:255:0 #yellow
 1.4 255:165:0 #orange
 2.0 255:75:0 #dark orange
 2.1 255:75:0 #dark orange
 2.2 124:252:0 #light green
 2.4 124:252:0 #light green
 4 139:37:0 #orange red
 8 139:37:0 #orange red
 9 199:21:133 #violet red
 40 199:21:133 #violet red
```

color_regrslope.txt

```
 -4 139 26 26 #firebrick4
 -1 205 51 51 #brown 4
 -0.5 238 99 99 #indianred2
 -0.1 255 64 64 #brown1
 -0.01 255 215 0 #gold1
 0 255 215 0 #gold1
 0.01 255 215 0 #gold1
 0.1 99 184 255 #steelblue1
 0.5 30 144 255 #dodgerblue1
 1 16 78 139 #dodgerblue4
 2 25 25 112 #midnightblue
 3 71 60 139 #slateblue4
```

color_regrcoefdet.txt

```
 0 white
 0.25 blue
 0.5 green
 0.75 yellow
 1 red
```

color_elevation_diff.txt

```
-12      0 0 0
-5      188 47 54
-1.5    251 0 13
-1.2    163 0 8
-1.0    163 0 8
-0.5    107 155 0
0.0     205 247 111
0.5     107 155 0
1.0     69 3 111
```

```
1.2    69 3 111
1.5    108 10 171
5      93 38 128
12     1 1 1
```